BOUNDARY-SCAN INTERCONNECT DIAGNOSIS

FRONTIERS IN ELECTRONIC TESTING

Consulting Editor
Vishwani D. Agrawal

Books in the series:

Essentials of Electronic Testing for Digital, Memory, and Mixed Signal VLSI Circuits
M.L. Bushnell, V.D. Agrawal
ISBN: 0-7923-7991-8
Analog and Mixed-Signal Boundary-Scan: A Guide to the IEEE 1149.4 Test Standard
A. Osseiran
ISBN: 0-7923-8686-8
Design for At-Speed Test, Diagnosis and Measurement
B. Nadeau-Dosti
ISBN: 0-79-8669-8
Delay Fault Testing for VLSI Circuits
A. Krsti_, K-T. Cheng
ISBN: 0-7923-8295-1
Research Perspectives and Case Studies in System Test and Diagnosis
J.W. Sheppard, W.R. Simpson
ISBN: 0-7923-8263-3
Formal Equivalence Checking and Design Debugging
S.-Y. Huang, K.-T. Cheng
ISBN: 0-7923-8184-X
On-Line Testing for VLSI
M. Nicolaidis, Y. Zorian
ISBN: 0-7923-8132-7
Defect Oriented Testing for CMOS Analog and Digital Circuits
M. Sachdev
ISBN: 0-7923-8083-5
Reasoning in Boolean Networks: Logic Synthesis and Verification Using Testing Techniques
W. Kunz, D. Stoffel
ISBN: 0-7923-9921-8
Introduction to I_{DDQ} Testing
S. Chakravarty, P.J. Thadikaran
ISBN: 0-7923-9945-5
Multi-Chip Module Test Strategies
Y. Zorian
ISBN: 0-7923-9920-X
Testing and Testable Design of High-Density Random-Access Memories
P. Mazumder, K. Chakraborty
ISBN: 0-7923-9782-7
From Contamination to Defects, Faults and Yield Loss
J.B. Khare, W. Maly
ISBN: 0-7923-9714-2
Efficient Branch and Bound Search with Applications to Computer-Aided Design
X.Chen, M.L. Bushnell
ISBN: 0-7923-9673-1
Testability Concepts for Digital ICs: The Macro Test Approach
F.P.M. Beenker, R.G. Bennetts, A.P. Thijssen
ISBN: 0-7923-9658-8
Economics of Electronic Design, Manufacture and Test
M. Abadir, A.P. Ambler
ISBN: 0-7923-9471-2
I_{DDQ} Testing of VLSI Circuits
R. Gulati, C. Hawkins
ISBN: 0-7923-9315-5

BOUNDARY-SCAN INTERCONNECT DIAGNOSIS

by

JOSÉ T. DE SOUSA
Technical University of Lisbon

and

PETER Y.K. CHEUNG
Imperial College of Science Technology and Medicine, University of London

KLUWER ACADEMIC PUBLISHERS
BOSTON / DORDRECHT / LONDON

A C.I.P. Catalogue record for this book is available from the Library of Congress.

ISBN 0-7923-7314-6

Published by Kluwer Academic Publishers,
P.O. Box 17, 3300 AA Dordrecht, The Netherlands.

Sold and distributed in North, Central and South America
by Kluwer Academic Publishers,
101 Philip Drive, Norwell, MA 02061, U.S.A.

In all other countries, sold and distributed
by Kluwer Academic Publishers,
P.O. Box 322, 3300 AH Dordrecht, The Netherlands.

Printed on acid-free paper

Printed in the Netherlands.

To Fernanda and Lai-Kit

Contents

Foreword

I worked with José de Sousa for almost two years at Bell Labs, and I know him as a bright researcher with an inquisitive mind that does not take anything for granted, a debater who argues passionately until his opponent proves to be right (or, actually more often, wrong!), and a funny guy to boot. But I know very well that being a great researcher is not a sufficient condition for being a good technical book writer - in fact, it is not even necessary - and now he has co-authored his first book and asked me to write a foreword for it!

My initial apprehension turned out to be unfounded, as this is an exceptionally well-written book, and it is a great intellectual pleasure to read it. The book is based on José's Ph.D. thesis and is co-authored with Peter Cheung, who was the thesis advisor. As I expected, José debates every claim extensively with himself, and both the pros and the cons of everything are thoroughly examined, and, of course, eventually he wins the argument! But the real winner is the reader, whose understanding of both problems and solutions is greatly enhanced in the process.

I assume you are a potential book buyer, and you read this while considering the difficult trade-off between price and value. Allow me to skip beating around the bush, and acknowledge that the purpose of a foreword is to gently nudge you toward the "buy" decision. Instead, here I'm trying to push you really hard in that direction. I do this because I am convinced that later you'll thank me for it.

Since you have picked up this book, you already know that the field of interconnect testing and diagnosis is very important today, when building a system-on-chip relies on interconnecting predesigned cores for which the tests are typically supplied by their manufacturer. Accurate diagnosis is essential if the defective circuit (IC or board) has to be repaired. You might be tempted to think that, if cores and ICs are designed using boundary scan or similar techniques that separate interconnect test and logic test, then interconnect test and diagnosis are trivial problems, because we have unrestricted controllability and observability for every net. But the book quickly dispels this view in its critical review of the existing methods, some of which are shown to provide ambiguous diagnosis for realistic fault models such as strong-driver shorts.

The authors' research contributions presented in the book include a set of efficient algorithms for generating diagnosis vectors using graph coloring and enhancing techniques. These algorithms have the original feature of being adjustable: they can be tuned to provide more or less diagnostic resolution. Compared with the existing methods, they can produce shorter diagnosis tests and/or better diagnostic resolution.

The book also proposes a framework for quantitative assessment of the diagnostic resolution of a test set. This evaluation is based on circuit layout and the fault statistics regarding density, clustering, and extension. This assessment method is much more advanced than the current evaluation techniques, and the authors present experimental results that illustrate the capabilities of the existing methods and highlight the improvements obtained with the proposed methods. Board layouts and statistics published for a real manufacturing process from HP have been used in the evaluations.

No matter what you are - researcher, designer, graduate student, or test engineer - this book will provide a lot of useful information to you.

Miron Abramovici
Bell Labs — Lucent Technologies
Murray Hill, New Jersey
November 17, 2000

Preface

This work started with a research project at Imperial College in collaboration with Schlumberger Technologies, and later gave rise to a doctoral dissertation. This book is based on that thesis to which we have added a more global perspective. Its focus is on test and diagnosis of wire interconnects in boundary-scan electronic assemblies (EAs), such as printed circuit boards (PCBs), multi-chip modules (MCMs) or systems on a chip (SOCs). The book covers interconnect fault modeling, test and diagnosis methods, and assessment of diagnostic resolution. Although boundary-scan is currently the main application of interconnect diagnosis, the methods presented here can be used in any environment where logic and interconnects are separated. Other applications include interconnects in field programmable logic arrays (FPGAs) and bare board test and diagnosis.

In the last three decades many interesting papers have been published on the subject of test and diagnosis of interconnects. This book is a synthesis of the existing methods and a report on the new techniques we have contributed to this technology. These methods use digital signals and a layout description of the EA for performing diagnosis. The electrical effects of deep-submicron interconnects are not covered in this research. Since interconnect circuits are one of the simplest electronic circuits that can be imagined, we have been able to provide a very comprehensive solution for the test and diagnosis problem. This study may be used as a departure point when addressing more complex circuits and systems.

For the diagnostic methods studied in this work we assume that the IEEE boundary-scan (BS) standard is being used to apply the tests to the EA. Although BS is the principal mechanism used for applying the schemes discussed in this work, this book is not about BS design and operation. The need for knowing the details of BS is fulfilled by the already available literature. The strength of this book is in methods for providing effective interconnect diagnosis patterns for BS. The BS architecture is only described to the extent of giving the user the necessary background to understand the contents of the book.

Three types of potential readers can be identified for this book: (1) researchers, (2) test engineers and (3) engineering managers. For researchers, innovative ideas on electronic test and diagnosis are presented, supported by a solid theoretical background. For test engineers, this book is a good reference guide to the various interconnect test and diagnosis schemes. Engineering managers may use the book to decide which is the best scheme for a given manufacturing process, and to learn about potential applications of BS.

Motivation and purpose

The last step in EA manufacturing is the final production test. An EA that passes the test is deemed defect-free and shipped to the customer; EAs that fail the test are deemed faulty, and diagnosis is performed. If diagnosis is successful the fault is repaired, unless the damage is so bad that it becomes cheaper to discard the EA. Most of the times, it is more viable to repair a faulty EA than to discard it. The fact that faulty EAs are not discarded is also beneficial for the environment.

Besides production test, diagnosis is also needed in maintenance testing. Maintenance tests are applied to EAs already in the field, on a regular basis or after they have failed. If an EA fails a maintenance test, diagnostic methods are applied to determine which faulty components have failed. Then, depending on the result of diagnosis, it may be decided to replace the faulty parts or the entire EA. Hence, diagnosis plays an important role in EA manufacturing and maintenance.

Testing and diagnosing EAs is a computationally difficult problem. The main difficulty of this problem lies in generating stimuli that will activate faults in parts not directly accessible from the EAs inputs, and will propagate the responses to one or more EA outputs. This is the classical controllability/observability problem of electronic testing.

A traditional solution for the EA controllability/observability problem has been the use of a technique called in-circuit testing (ICT). However, with EAs becoming more and more miniaturized, ICT has become obsolete. In its replacement came the BS technique, inspired by the *scan-design* methodology, developed to ease the same kind of problem at the integrated circuit (IC) level. To facilitate test and diagnosis of EAs, it has been decided to make BS a standard, which can be used by any chip manufacturer. In this way, EAs containing only BS chips can universally be tested and diagnosed. A working group called the Joint Test Action Group (JTAG) was formed to create the Standard Access Port and Boundary-Scan Architecture, which became the IEEE 1149.1 standard in 1990. This standard is now a widely accepted test protocol, whose benefits are beyond dispute.

The most important use of BS is testing and diagnosing faults in wire interconnects between ICs, introduced during the assembly process. Hence the subject of this book. The methods contributed by this research can drastically reduce the number of test patterns required for interconnect diagnosis, and offer extremely accurate physical level diagnostic information.

Some critics may say that the gains in interconnect diagnosis that these methods achieve are hidden by the costs of test sets for BS non compliant components and component clusters. In fact, when BS non compliant components exist on the EA, a significant portion of the test resources are spent to test the interconnects between these components, and, in some cases, the components themselves. This is a fair criticism for the time being. However, for reasons that will be explained below, the use of BS non compliant components is in decline, and will tend to disappear. In a scenario where only BS compliant ICs are present on the EAs, the proposed methods can achieve substantial gains.

Another criticism that we have faced sometimes says that the existing interconnect diagnosis methods already produce test sets of an adequate size and diagnostic capability. In our view, this is a short sighted comment. First of all, in terms of diagnostic resolution, the current schemes are very poor due to the lack of layout level information. In the current approaches, physical information is introduced only later during the fault isolation and repair phases, which is less efficient. As for the size of the test sets, the criticism is similar to saying that mobile telephones are already small enough. Mobile phones are getting smaller and cheaper every day, which enables more functionalities to be added to them. For example, video mobile telephony will be a reality soon. In the same way, if interconnect test sets get more compact, it becomes possible to think of online test and diagnosis based on stored patterns, wireless testing, etc. A future development suggested in this book is *EA automatic repair technology*, only possible with the accurate physical level diagnostic information methods that we have developed.

BS technology also enables testing the ICs already mounted on the EA. However, this functionality is not commonly used in production test because it is expensive and in most cases unnecessary. In fact, it may take a significant amount of time to test an IC using its BS infrastructure. Also, since the ICs have been individually tested after they were fabricated, and since it is unlikely that they have become faulty during the assembly process, IC testing during production test can be dispensed with. On the other hand, during maintenance test, testing individual ICs becomes relevant. For a system already in the field, it is more likely that its ICs will fail than its wire interconnects. Note that, due to their high reliability, wire interconnects which already passed production test will normally remain fault-free. Thus, during maintenance test the ICs are the important parts to test.

The evolution in IC test methodologies suggests that a great proportion of the future ICs will also incorporate built-in self-test (BIST). With BIST, the ICs are able to test themselves, and thus no external tests applied using the BS architecture are necessary. In fact, the BS architecture already incorporates an instruction to initiate self tests in the ICs equipped with BIST. Hence, when finding a faulty chip, several self tests can be initiated simultaneously in several ICs. The results of the self tests are then collected serially via BS and shifted out. With BIST, locating a faulty chip in a faulty EA becomes much more efficient.

This work is based on the scenario that practically all chips will incorporate BS or some future scheme derived from it. We believe that this will be a reality soon, but before that a few problems have yet to be solved, and are actually being solved by other researchers. Although those problems are out of the scope of this research, it is worth enumerating the main ones: (1) some ICs are difficult to equip with BS; (2) glue logic parts are still required in many EAs; (3) EAs might include analog and mixed-signal components, for which entirely different considerations apply. In the following paragraphs we will elaborate on each of these problems.

Some chips, such as random access memories (RAMs) do not normally incorporate BS. RAMs are compact and dense regular structures, which cannot tolerate an area increase or performance loss. In these respects, BS would represent an unbearable overhead. More examples of ICs that do not incorporate BS, for other technical or historical reasons, could be given. Some of these chips will overcome these problems and incorporate BS in future versions. Others will still resist BS introduction and will find a clever work around. One assumption currently used in alternative solutions is that the non compliant chip only connects to BS chips. Thus, the non compliant chip and its interconnects are entirely tested by the BS structure of its neighbor chips.

Connecting two ICs to each other is not a trivial problem, and many times additional smaller components must be used to properly interface the ICs. These components are usually known as glue logic, and, due to their low complexity, are not usually BS compliant. Glue logic components represent a problem for BS testing, but, fortunately, their use is declining. Just open the box of a modern electronic system and compare it to, say, a fifteen year old system. In the old system many small active components (e.g., the 74000 series) can be seen on the boards, which contain few large ICs. In the modern system, many more large ICs can be seen, with a lot less glue logic components around. Several factors contribute to the decline in the use of glue logic components. Among the most important, we will mention (1) the standardization of interfaces and (2) the rise of configurable logic technology.

Standard interfaces between ICs introduce plug-and-play capability in EA development, and dispense with the use of glue logic components. One example of a standard interface is the PCI bus, widely used in computer systems. Many subsystems (boards for various purposes) have their own PCI bus for interconnecting their chips. A PCI bus of a subsystem can be connected to the PCI bus of the main system (or to the PCI bus of another subsystem) by means of a PCI bridge IC.

Configurable logic technology, field programmable logic arrays (FPGAs) and complex programmable logic devices (CPLDs), are replacing traditional ICs, due to their flexibility and amazingly short application development time. Together with the target application, all the logic components necessary to interface to other ICs can be included in a configurable chip, completely eliminating the need for glue logic components. Also, if one wishes to upgrade the system by replacing some IC a with newer version, the interface to the new IC can be obtained by simply reprogramming the configurable logic accordingly. It is likely that in the near future hybrid chips consisting of an IC core surrounded by programmable logic will emerge. Such technology would make chip to chip interfaces even easier to implement. In terms of test and diagnosis, EAs containing only configurable logic chips can be fully tested with BS, as FPGAs and CPLDs normally include BS.

The problem of mixed-signal and analog ICs is being tackled on a few different fronts. An extension of BS to analog and mixed-signal devices, the IEEE 1149.4 standard, has recently been approved. With this new standard, it becomes possible to control and observe the pins of analog and mixed-signal ICs in a similar way to BS in digital ICs. Additionally, as the technology of integrated passive components (resistors, capacitors and inductors) evolves, the use of external discrete components will also decline. The test of discrete components is known as parametric test, [1] and consumes a significant part of the testing resources. Therefore, the impact of integrated passive component technology will be very positive, since there will be few or no discrete components to test and diagnose. This technology is also beneficial to greater miniaturization, better performance, etc.

A new methodology

In the context just described, the contributions of the present work are: (1) *interconnect fault modeling* is put into a new perspective that distinguishes between synthesis of diagnostic stimuli and analysis of responses; (2) a novel and formal analysis of traditional interconnect diagnosis schemes, usually known as *behavioral interconnect diagnosis schemes*, is presented; (3) algorithms for *structural interconnect diagnosis* have been developed; (4) an in-depth study of *diagnostic resolution assessment* is presented. More information on each of these contributions is provided in the next paragraphs, respectively.

In traditional electronic testing, fault modeling provides target fault behaviors to guide automatic test pattern generation. It is not always as important that these models are close to the actual fault behaviors, as long as the tests guarantee rejection of faulty circuits. In this work, it has been realized that in diagnosis two uses of fault modeling can be identified: fault models are used to generate diagnostic stimuli and to analyze faulty responses. The fault models for these two purposes need not be the same. When generating diagnostic stimuli the fault models should be simple, in order to reduce the complexity of test generation. When analyzing faulty responses, the fault models should be more general, so that fault behaviors not previously considered can be accounted for. We have called the first type of fault models the *synthetic* fault models, and the second type of fault models the *analytic* fault models. The fault models commonly referred to in the literature, i.e., the stuck-at fault, the wired-and

[1] Parametric tests verify if a discrete analogue component is characterized by the correct parameters (resistance, gain, etc.), which may be wrong or out of range due to process deviations or errors during component placement.

and wired-or short-circuit faults are synthetic fault models. The concept of analytic fault models is new and is introduced for the first time. It originates from the realization that synthetic fault models are inadequate for diagnostic analysis (we show they may lead to erroneous diagnoses or, in some cases, to no diagnoses at all). The commonly used synthetic fault models are used in this work only to generate test sets. The responses are analyzed with the new analytic fault models, providing better diagnostic information.

Behavioral schemes have been very popular in interconnect fault diagnosis. The main advantage of behavioral schemes is their independence from the physical layout. They can be developed at the netlist level, and remain valid despite layout modifications. However, they account for more faults than the ones that can actually occur, which affects their ability to generate compact test sets and reduces their diagnostic resolution. In this work, we formally analyze almost all the behavioral schemes, proving their diagnostic properties with respect to the synthetic fault models used. We have also derived a new scheme, which we called the *N sequence*. This scheme is interesting from the theoretical point view because it is the shortest test set that possesses a property called the diagonal independence property. This book presents a rigorous and comprehensive study of behavioral schemes for interconnect diagnosis.

Unlike behavioral schemes, structural schemes do exploit layout and defect information to further optimize the size and capability of test sets. In fact, layout and defect information can considerably constrain the space of physically possible faults, causing the number of test vectors to drop by up to an order of magnitude. Structural schemes for detecting interconnect faults based on the graph coloring problem have long been known, but the possibility of extending them to diagnosis had not been fully developed. In this research we show that structural schemes can be used not only to reduce the number of diagnosis test vectors, but also to improve diagnostic resolution. We show that the number of diagnosis vectors can be reduced by about one order of magnitude, while actually improving the diagnostic resolution. Our method is called *statistical color mixing*, and is based on combining defect statistical information with a graph technique. With structural diagnosis, diagnostic information can be given at the physical level, which is extremely useful during the fault isolation and repair phases. Physical level diagnostic information also creates the opportunity for automatic repair technology. Since the defects can be accurately located, we can think of developing repair robots that will use that information to find and repair the defects. Like for the behavioral schemes, we present a formal and detailed analysis of the properties of the structural diagnostic schemes.

Proving the formal properties of diagnostic schemes helps in establishing levels of qualitative diagnostic capability, but does not provide a quantitative evaluation of diagnostic resolution. That is only possible with statistical information on the characteristics of real defects: extension, topology, etc. Based on defect statistical information, we have proposed a method for assessing diagnostic resolution prior to test set application. With such a method, we can select the best diagnostic scheme for a statistically characterized process. Furthermore, if the process statistical parameters change over time, the diagnostic scheme can be dynamically updated to reflect the change.

Considerable experimental data is presented in support of our contributions. We have implemented a tool for fault extraction, a tool for generating the various behavioral and structural diagnostic schemes, and a statistical diagnosis simulator. In our experiments, actual PCB layouts have been used, and the simulation environment has been statistically characterized with parameters from a real process.

Structure of the book

This book is organized in seven chapters.

An introduction to boundary-scan test and diagnosis of interconnects is given in Chapter 1. This includes an explanation of defects in EAs, test and diagnosis practices, and in-circuit testing. The other standards of the family, IEEE 1149.4 and 1149.5 are quickly explained. The chapter provides a discussion of the problem of interconnect defects, and outlines the contributions of this research.

Chapter 2 presents a model for an interconnect circuit and models for interconnect faults. The distinction between synthetic and analytic fault models is introduced.

In Chapter 3, an original formal analysis of behavioral methods for interconnect diagnosis is presented, and the scheme developed in this research, the N sequence, is explained. At the end of the chapter, a table summarizing the properties of all the schemes analyzed is given.

Structural methods are explained in Chapter 4. Schemes by other authors are formally analyzed, and the scheme proposed in this research, the statistical color mixing scheme, is described. Three appendices support the materials presented in Chapter 4. Appendix A gives the syntax of the layout file format; Appendix B and Appendix C describe the graph coloring and graph enhancing algorithms, respectively, used in structural diagnosis.

In Chapter 5, a statistical framework for assessing the diagnostic resolution of generic test sets is proposed.

Experimental results are presented in Chapter 6. Results on fault extraction, test vector length and diagnostic resolution are discussed. These results show the advantages of the schemes proposed in this research, when compared to schemes by other authors. The computation time aspect of various algorithms is studied.

Chapter 7 concludes the book. A summary of the covered material is given, conclusions are drawn, and directions for future work are pointed out.

This book uses many different acronyms, abbreviations and mathematical symbols to represent different entities and quantities. These symbols are explained the first time they appear in the text, and have also been listed in two unnumbered appendices in the last pages of the book.

Acknowledgments

This work would not have been possible without the help of several people and institutions.

We would like to thank the following people: David Shen, who is now with Fujitsu, for his involvement in the project during his one year postdoctoral appointment at Imperial College in 1995; Vishwani Agrawal from Bell Labs, who reviewed the manuscript and made several useful comments and suggestions to the text; Herman Desmier and Dave Webber from Schlumberger Technologies, for providing an industrial perspective on the work; Will Moore from the University of Oxford, for his comments and discussions about particular aspects of the work; Miron Abramovici from Bell Labs, for encouragement and expert advice in the field of diagnosis; James Finlay and Cindy Lufting from Kluwer Academic Publishers, for their friendliness and professionalism.

The following organizations are acknowledged: the Imperial College of Science Technology and Medicine of the University of London, where this work was developed; Schlumberger Technologies in Ferndown, England, for industrial partnership; INESC/IST, Technical University of Lisbon, for partial sponsorship and support; the Foundation of Science and Technology of Portugal, for partial sponsorship; Bell Laboratories in Murray Hill, New Jersey, for support while part of the book was being written.

Finally, we thank our families, especially our wives Fernanda and Lai-Kit, for their patience and support while this book was being written.

1 INTRODUCTION

The focus of this book is on test and diagnosis (T&D) of wire interconnects using boundary-scan electronic assemblies (BSEAs). BSEAs are normally printed circuit boards (PCBs), multi-chip modules (MCMs), or systems on a chip (SOCs). Besides BSEAs, the techniques presented here are applicable to bare board test and diagnosis, and can be extended to deal with interconnects of Field Programmable Gate Arrays (FPGAs). The book is a synthesis of three decades of work on interconnect T&D, and adds its own perspective and contributions to this technology. Wire interconnect signals are treated at the logical level; it should be clear that the problems specific to deep-submicron interconnects are not covered.

This chapter explains why interconnect faults are important, describes the techniques currently used for wire interconnect test and diagnosis (WITD) and gives an overview of the main contributions of the research that has given origin to this book.

This introduction is organized as follows. In Section 1.1, potential readers of this book are identified. Electronic assemblies (EAs) and their defects are discussed in Section 1.2. The process of testing and diagnosing EAs is outlined in Section 1.3. The possibilities for automatic repair are discussed in Section 1.4. In Section 1.5, in-circuit testing (ICT), the predecessor of boundary-scan (BS), is explained. In Section 1.6, BS, or the IEEE 1149.1 standard, is examined in terms of its architecture and operation. Other related standards, the IEEE 1149.4 for testing of analog and mixed-signal EAs, and the IEEE 1149.5 for system test, are outlined in Section 1.7. In Section 1.8, the central problem of this work, the WITD problem, is stated. The contributions of this research are outlined in Section 1.8.2, and, finally, a summary of the chapter is given in Section 1.9.

1.1 Who this book is written for

This book targets three types of readers: (1) researchers who are interested in the WITD problem, (2) test engineers who want to implement WITD schemes and learn about their features, and (3) engineering managers who need to select a T&D methodology for interconnect faults.

For researchers, this book presents the theoretical issues in great detail. The most important schemes and their main properties are formally proven. A set of innovative ideas on fault modeling, generation of diagnostic vectors for faults extracted from the layout, and statistical assessment of diagnostic resolution are presented. These are interesting problems in the area of electronic testing, which transcend the central theme of this work, the WITD problem.

For test engineers and managers, the book provides a comprehensive survey of WITD methods. The theoretical aspects of the different methods are less interesting for these readers. Nevertheless, they will appreciate the fact that the important properties of the various methods are systematically enumerated, which they can survey without getting into the intricacies of formal analyses. Managers may also find interesting the perspective that this book offers on the WITD problem in particular, and about BS in general.

1.2 Electronic assemblies and defects

Modern EAs are complex and may have tens of thousands of interconnect nets. It is either impossible or too expensive to fabricate EAs without defects. Some production defects can cause *faults*, i.e., cause the EA to behave in a way that violates its specification.

The fact that faulty EAs may be produced leads to the concept of *production yield* (Y), defined as the fraction of fault-free units after manufacture, but before test. There are no typical values for Y; it may vary from process to process, and for different designs in the same process. Some EAs might be manufactured with a yield as high as 90%, whereas other designs may hardly manage a 20% yield. However, within the same process, Y decreases with design complexity. [1] It is during the *production test* that good EAs are separated from the faulty ones.

Doubtlessly, the principal cause of low yields is defects introduced during the manufacturing process. There is, however, another non-negligible cause of low yields: some EA components may already be faulty before being assembled. These faulty components are called *escapes*. For a particular component, the fraction of escapes with respect to the whole production lot is called the com-

[1] Larger the number of components in the EA, more likely it is that something will go wrong.

ponent's *reject rate* or *defect level* (DL) (Williams and Brown, 1981; Agrawal et al., 1982). It is important to control DL, or otherwise the product will lose competitiveness (Sousa et al., 1996a). Particularly, complex integrated circuits (ICs) are extremely defect-sensitive. They have to undergo complex test procedures, in order to ensure a low DL. In some cases, the cost of testing becomes the main manufacturing cost. It is important that the components used in EA manufacturing have a low DL. For small component DLs, and neglecting higher order terms, the overall DL of the EA is roughly the sum of the individual DLs. Some EA manufacturers even perform incoming component inspection, in order to enforce a low enough component DL.

Testing an EA by trying to exercise all its functions may not be a practical approach. EAs can be quite complex, and such an *exhaustive* or *behavioral* kind of test could take forever to apply. An alternative is the so called *structural* test; a set of possible faults is extracted from the layout, and a test set is derived targeting the detection of the extracted faults.

Different types of production faults can occur in EAs. According to Johnson, a typical distribution of faults in PCBs manufactured with surface mount technology (SMT) is as reproduced in Table 1.1 (Johnson, 1993). From this table one can see that *fine pitch shorts and opens* (type I) is the dominant fault type, with 80% incidence. These faults are caused by interconnect short circuit and open circuit defects, affecting the solder joints in SMT components (Wassink, 1989). SMT components have a very fine *pitch* (pin spacing), which in this study was 20-25 *mil* (1 *mil* = 1/1000 of an inch). This makes pin solder joints very susceptible to defects introduced during the soldering process.

Table 1.1. Typical fault distribution on SMT-based PCBs (Johnson, 1993).

Fault type	*Description*	*Incidence (%)*
I	Fine pitch shorts and opens	80
II	Device faults	11
III	Other shorts and opens	7
IV	Structural faults	2
Total		100

The next most important fault type is *device faults* (type II). Its occurrence rate of 11% is much lower than the 80% incidence of type I. Type II faults are usually due to escapes, damage during the assembly process, or infant mortality. The next fault type is *other shorts and opens* (type III). They are caused by interconnect defects at locations other than the component pins, and can be due to bare PCB escapes, or damage during the assembly process. Their low occurrence rate is due to bare board pre-testing (Chen and Hwang, 1989; Leung, 1993; Vaucher and Balme, 1993), a process that practically eliminates bare board defects. Damage during the assembly process occurs very rarely in a mature process. The last fault type is ***structural faults*** (type IV), which is a missing, misplaced or wrongly assembled device. Type IV faults are the most unlikely ones as evidenced by their lowest occurrence rate (2%).

Looking at the fault spectrum above one concludes that 87% of the faults (types I and III) occurring in this SMT process are due to wire interconnect defects. Interconnect defects are also the dominant cause of faults in MCM technology (Johnson, 1993). These observations illustrate the importance of interconnect defects.

1.3 The test, diagnosis and repair process

When manufacturing discrete components or monolithic ICs, it is common practice to discard the ones that have not passed production test. Repairing discrete components and ICs is out of the question, both for technical and economic reasons. On the other hand, repairing EAs is often technically and economically viable. Before repairing a faulty EA, the defects or failures causing the fault must be located or, in other words, *diagnosis* must be performed. Hence, EAs have to be *tested*, not only to check whether they are working, but also to be *diagnosed*, so that their faults are identified and repaired.

The test and diagnosis process is carried out using automatic test equipment (ATE) and may be described as illustrated in Figure 1.1. Test preparation is performed on a workstation by writing a *test program* for the circuit under test (CUT). The test program is compiled to produce the *test data*. The test data consists of the stimuli to be applied to the CUT, and the expected responses. Test programs must be *debugged* before they can be used. To do this, the test program is uploaded onto the tester, and applied to EAs that contain known faults. The objective is ensuring that real faults can be successfully diagnosed with the newly developed test program. Once debugged, the test program becomes ready to be used in production test. During production test the test program becomes resident in the tester, and is run on all fabricated EAs. Test stimuli are applied to the CUT and the actual responses are compared with the expected ones. If the actual responses match the expected ones, the CUT passes the test and is ready to be shipped. Otherwise, the response is sent back to the control workstation, which produces possible diagnoses. The diagnoses are passed to the repair technician who isolates the real faults, and carries out repairs at the repair station. Repaired circuits have to be tested again, as repairing might be insufficient and/or new faults might be introduced.

The unitary cost CU (cost per EA) of the test, diagnosis and repair step is now considered. A simple cost model is given by the following equation:

$$CU = CATE + CTA + (1 - Y)(CDI + CRR), \tag{1.1}$$

where $CATE$ is the unitary cost of the ATE system, CTA is the unitary cost of test application, CDI is the unitary cost of diagnosis and fault isolation and CRR is the unitary cost of repair and retest. $CATE$ is a function of the capacity and quality of the ATE system. CTA depends on the time for mechanical handling of the EA (putting it on and off the tester) and electronic application

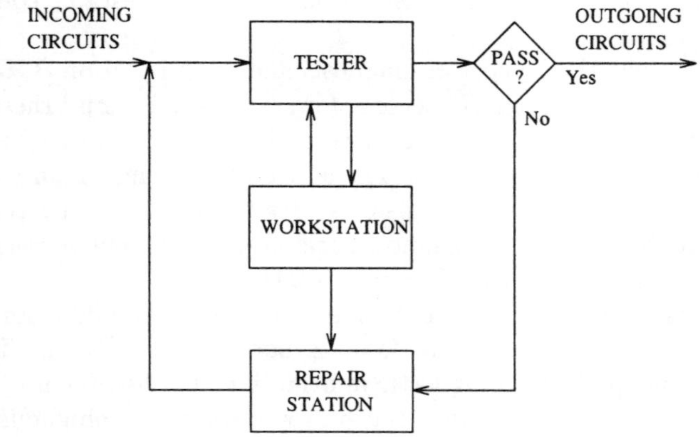

Figure 1.1. The test, diagnosis and repair process.

of test data. *CDI* is proportional to the diagnosis and fault isolation time. *CRR* depends on the repair time, cost of replaced items and reapplication of tests. *CDI* and *CRR* are multiplied by $(1 - Y)$ because these costs only apply to faulty EAs.

To optimize the overall *CU* cost, the context should carefully be considered. For low production volumes it is very important to minimize the cost *CATE*. This can be achieved by minimizing the amount of test data, so that one can buy an ATE system with little memory and low speed. Also, a cheap ATE system does not need to have mechanical features for handling EAs, which can represent a significant investment if the production volume is low.

For high production volumes the price of the ATE system becomes less important, but it is still significant. An ATE system for high production volumes needs good mechanical features for fast and safe handling of the EAs. An ATE system with such features will contribute to reduce the cost of test application *CTA*. If, additionally, the ATE system needs a large memory capacity and a high speed, the cost can rapidly become prohibitive. One way of mitigating this cost is to consider techniques for minimizing test data size.

From the previous two paragraphs, it is concluded that for both low and high production volumes, minimizing the test data is always an issue. Since, as seen in Section 1.2, interconnect faults are the most likely faults, an important portion of the resources should be spent to ensure that these faults are tested. This provides motivation to study methods for testing and diagnosing interconnects, which simultaneously yield effective and compact test data.

For device and structural faults (types II and IV in Table 1.1), less T&D effort is needed, since their occurrence rate is lower. Testing chips already mounted on the EA is expensive, and should be avoided. One alternative solution is making sure that the chips have a low DL, so that testing can be dispensed with. This can be done by choosing a high quality supplier or performing incoming component inspection. Using self-testable components is another possibility.

For a high yield Y, the cost of diagnosis and fault isolation (CDI) and the cost of repair and retest (CRR) become lower. This is because these terms are affected by the factor $(1 - Y)$ in Equation (1.1). On the other hand, if Y is low, these costs become important. The cost CDI depends mainly on the cost of fault isolation, since diagnoses can usually be issued fast by the diagnosis software. The fact that these diagnoses are often ambiguous impacts the fault isolation time, and consequently the cost CDI.

The fault isolation time can be viewed as the time interval from the arrival of diagnostic information until the fault to be repaired is found. This time is a function of the quality of diagnostic information. Precise diagnostic information corresponds to a short fault isolation time, whereas ambiguous diagnostic information leads to a long fault isolation time. Since repair costs are usually high, fault isolation time represents an important economic factor.

Various techniques have been used to improve fault isolation time (Bateson, 1985). Computer-aided repair (CAR) systems first came into use in 1976 as what was called a *paperless repair system.* With such a system the repair technician would enter the ID of the faulty EA in a terminal, and a message containing the diagnostic information would be displayed. Because textual diagnostic information is slow to interpret, in 1981, an ATE company introduced graphical diagnostic information. The technician would see on a screen the components to be replaced and/or pads to be unsoldered. However, graphical diagnostic information can be efficient only if the diagnostic information is accurate enough. Otherwise the technician could be looking for faults that do not exist and missing the ones that really are there. The issue of fault isolation time is so important that some CAR systems use light beams to pinpoint faulty components, pads or polarities directly on the board, so that the repair technician does not get distracted looking at other locations on the EA.

With *perfect* diagnostic information the fault isolation time can be reduced to a minimum, so that CDI becomes CDI_{min}. Thence, it is convenient to define *fault isolation efficiency* FIE as

$$FIE = \frac{CDI_{min}}{CDI}. \tag{1.2}$$

In this work *diagnostic resolution* (DR) is defined as being the ratio of the number of actual defects or failures in a faulty EA by the number of defects or failures diagnosed by the diagnostic method being used. It will be assumed that FIE can be equated to DR:

$$DR = FIE. \tag{1.3}$$

It is also being assumed that the set of actual defects or failures is always a subset of the set of diagnosed defects or failures. As will be seen, this is true for interconnect faults, and means that $0 < DR \leq 1$. Combining the last two equations, it is possible to write

$$CDI = \frac{CDI_{min}}{DR}. \tag{1.4}$$

The equation above clearly shows that to minimize the cost CDI one needs to maximize the diagnostic resolution DR. Maximizing DR is major concern in this work.

1.4 Automatic repair

If repair costs are so important, then automating the repair process is certainly something that would be well received by the industry. As will be explained, accurate diagnostic resolution is the key to *automatic repair*.

Two qualities of diagnostic information are essential for the development of automatic repair techniques: *completeness* and *high resolution*. A diagnosis is *complete* if the set of diagnosed items is a superset of the set of faulty items. A diagnosis is of *high resolution* if the set of diagnosed items closely matches the set of faulty items. If the diagnostic information possesses these two qualities, and the necessary robot technology can be deployed, then the conditions for automatic repair will have been met.

Note that, with automatic repair, all diagnosed parts will be replaced and their interconnects repaired. If the number of diagnosed items is significantly greater than the number of faulty items, automatic repair can become exceedingly expensive. In other words, if DR is low, a considerable number of fault free parts and interconnects will be unnecessarily repaired. Hence, it is important to maximize DR in order to minimize the number of unnecessary repairs. Note that, if the yield is low (and it often is in complex EAs), a large number of EAs will need repair. Whence, *diagnostic resolution becomes a decisive factor in terms of automatic repair cost effectiveness.*

1.5 In-circuit testing

In-circuit testing (ICT) is a test application technique which has been widely used in PCBs (Bateson, 1985; Buckroyd, 1994; Crook, 1990; Crook, 1979). With ICT, test stimuli and responses are applied and collected directly from the nets using a *bed-of-nails* fixture. Such a fixture is illustrated in Figure 1.2.

The fixture's test pins touch the board surface at points where electrical contact is possible, for example, at component pins or at pads specifically designed for that purpose. ICT is a powerful technique capable of testing digital as well as analog EAs. In fact, test pins can apply or read both digital and analog signals, and diagnosis is greatly simplified by the fact that internal nets ca be accessed directly.

ICT can perform fast interconnect testing, and is also used to test discrete components in the EA. The testing of components consists of checking if they have been correctly mounted, verifying if the passive analog components have their parameters within an acceptable range, etc. With ICT, components in the EA can be tested individually, but, as mentioned in the previous section,

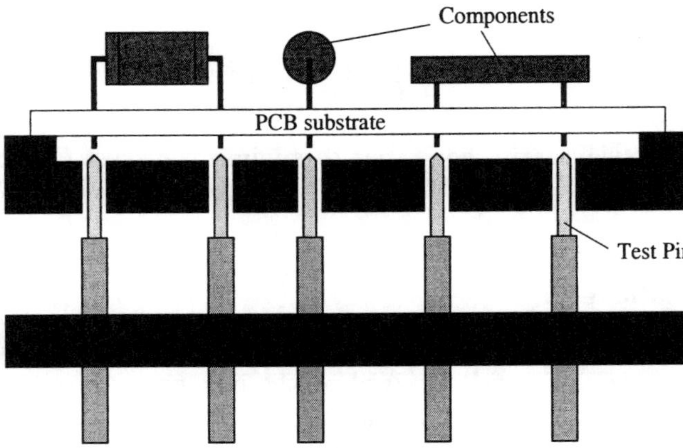

Figure 1.2. A bed-of-nails tester.

that should be avoided to save time and money. Also, tests should be applied quickly because of the *backdriving* problem: while driving a net with a test pin, the normal driver of the net (i.e., the output of some chip in the board) is backdriven, and may burn if the action takes an excessive time.

In spite of all its advantages, ICT is becoming obsolete as component densities increase in PCBs (Bleeker et al., 1993; Maunder and Tulloss, 1990; Parker, 1992). SMT chips can be highly complex and miniaturized, and come in small packages with very low pitches. Such low spacings between the IC pins make it difficult and expensive to miniaturize bed-of-nails fixtures accordingly. For technologies that enable direct mounting of silicon dies on *both sides* of the board, such as tape automated bonding (TAB) or chip on board (COB), this problem becomes even worse.

1.6 Boundary-scan testing

ICT is rapidly becoming obsolete, and a substitute technology has become a major priority. In the 80s, a technology able to fill in this gap began to emerge. This technology was inspired by IC design for testability methods (Williams, 1982) and was called boundary-scan (BS) testing (Maunder and Beenker, 1987). In 1990, BS became an IEEE standard: the IEEE 1149.1 Standard Access Port and Boundary-Scan Architecture (IEEE, 1990) describes testability hardware for digital ICs to facilitate post-assembly T&D and system maintenance. With BS, EAs can be tested and diagnosed using dedicated external I/Os, with no need for direct access to internal nets.

1.6.1 Boundary-scan architecture

The BS architecture is shown in Figure 1.3. Four terminals, named Test Data In (TDI), Test Data Out (TDO), Test Clock (TCK) and Test Mode Select (TMS), are added to the EA specifically for test purposes. The same test terminals are

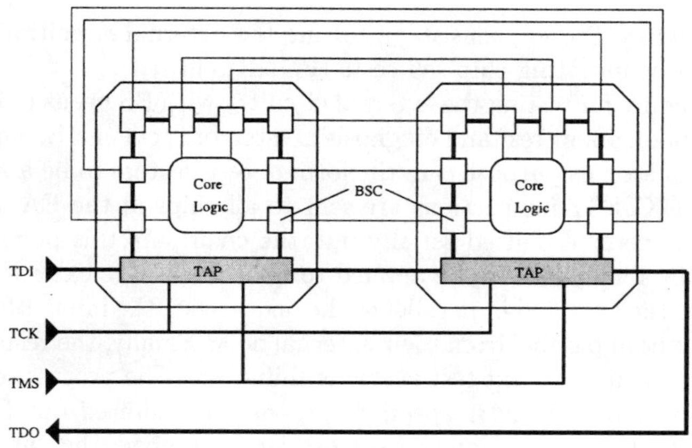

Figure 1.3. The boundary-scan architecture.

present in each IC on the EA. Each IC is provided with a controller, called the Test Access Port (TAP) controller, and with boundary-scan cells (BSCs) associated with the I/O pins of the IC.

Each BSC consists of a 1-bit shift register stage. Inside the IC, the BSCs are connected in series, so that a shift register known as the Boundary-Scan Register (BSR) is formed. In this way, the test data can circulate through all signal pins of the IC, solely using the ICs TDI and TDO serial I/Os. A BSC associated with an input pin can monitor the logic level present at the respective external net, or drive the respective input of the internal core logic. A BSC associated with an output pin can monitor the respective output of the internal core logic or drive the respective external net. Bidirectional pins can use special BSCs, which may both drive and monitor the respective external net or internal core logic terminal.

The ICs in the BSEA are connected in series, so that the test data can circulate through all chips, simply using the BSEA's TDI serial input and the TDO serial output. The signals TMS and TCK connect to all ICs in parallel. Optionally, there can be multiple BS chains in the BSEA.

1.6.2 Boundary-scan operation

The operation of the BS scheme is synchronized by the clock signal TCK and controlled by the signal TMS. In the TAP controller there is an instruction register (IR) to receive instructions, which determine what each IC is to do next. There are several instructions defined by the standard: EXTEST, the instruction for testing the external circuitry, INTEST, the instruction for testing the internal core logic, RUNBIST, for running self-tests in chips equipped with built-in self-test (BIST), and others. The INTEST facility is not commonly used in post-manufacture test. As with ICT, testing each chip in the EA individually has a severe penalty in terms of test time. Whence, a typical post-manufacture BS test consists of testing the external circuitry (EXTEST) and, optionally, initiating chip self-tests (RUNBIST).

The procedure for testing the external circuitry with BS needs to be clarified, for this is a book about test and diagnosis of interconnects in a BS environment. The test data for this purpose is in the form of test vectors to be applied to the CUT nets. EXTEST instructions are sent to all chips in the EA, after which the first test vector is shifted serially into the chain. At this point all output BSCs contain a logic level to be applied to their respective external nets. The test vector is then applied in parallel to the nets. Next, the input BSCs load the response vector in parallel from their external nets. Finally, the response vector is shifted out while the next test vector is shifted in. The process repeats until all vectors are applied and respective responses are shifted out for analysis. Note, however, that before any test vectors can be applied, the integrity of the BS infrastructure must be verified. Without this step any test results might be invalidated. Methods for performing BS integrity tests are discussed in the literature (de Jong, 1991).

BS can also be used for *maintenance test.* Maintenance test is different from post-manufacture test, since the reliability failure spectrum significantly differs from the post-assembly fault spectrum given in Table 1.1. The most likely fault cause during maintenance test is a chip failure. In terms of reliability, failing chips are much more likely than interconnect faults. In post-manufacture test it is the other way round: defective interconnects are the main fault cause. To identify the failing chips, the ICs have to be tested individually, which can be accomplished either by the INTEST or the RUNBIST facility. For the latter, the chip must be self-testable. Another difference between maintenance test and post-manufacture test is the fact that in maintenance test the effectiveness of diagnosis is more important than its efficiency.

Boundary-scan has been initially devised to address the testing problems of EAs. Now, it is also beginning to be used for unforeseen applications such as in-situ reconfiguration of programmable devices such as field programmable gate arrays (FPGAs) and complex programmable logic devices (CPLDs). The internal access provided by BS is a powerful tool, and perhaps more uses for BS will be devised in the future.

1.6.3 Test data serialization

With BS, T&D can be performed solely using the BS dedicated I/Os, reducing the complexity and cost of the ATE system. In fact, a simple personal computer (PC) with a BS interface can be used. However, with such a simple test equipment, large BSEAs would take a long time to test, which would make them expensive. The main reason is the need to perform test vector *serialization.* This operation reformats the test vectors in order to obtain the serial bitstreams that are applied by the BS infrastructure. There are several options for serialization (Bleeker et al., 1993):

1. *Run-time software serialization.* A software program produces bitstreams from the parallel test data, which are serially applied as they are being

output. This option saves the storage requirement but increases the test time several times. Note that the serialization program runs much slower than the BS hardware is able to operate.

2. *One-time serialization with disk storage.* Bitstreams are produced once and stored in the tester's hard disk. Then, a bitstream is accessed from the disk for every new assembly being tested. This is still a slow process since some BSEAs might require bitstreams of several hundred megabits.

3. *One-time serialization with buffer storage.* Bitstreams are stored in a hardware buffer and applied from there. In this case the BS hardware can run as fast as it can. However, for large BSEAs, such buffering capabilities can only be offered by expensive ATE systems, or hardware especially designed for that purpose.

4. *Run-time hardware serialization.* Dedicated hardware executes the whole serialization process and applies the test at the BS hardware maximum rate. This solution is no doubt a good one, but a tester capable of implementing it will be significantly more expensive.

In conclusion, whatever the ATE solution for BS is, it has advantages and disadvantages. In any case, it is important to minimize the amount of test data, in order to reduce test time and ATE cost. If the test is compact, BS self-test may be an alternative worth considering (Jarwala and Yau, 1991; Lubaszewski and Courtois, 1992). With BS self-test, the BSEA either stores or generates its own tests, and applies them to itself using the BS structure, avoiding the costs of external test application.

1.6.4 Applying power to shorts

Similar to the backdriving problem in ICT, in BS the problem is how long can the BSCs drive a low impedance short-circuit before disruption occurs. This is difficult to know, but sometimes ICs do burn because of that. The problem just mentioned provides extra motivation for reducing the size of test sets. Reducing the number of test patterns reduces the time drivers are subjected to strong currents, minimizing the probability of disruption.

1.6.5 Adaptive test and diagnosis

One solution that has been proposed for further minimizing the size of test sets is the so called adaptive T&D solution (Jarwala and Yau, 1989). According to this approach a set of minimal tests, aiming solely at fault detection, is applied first. If the circuit passes this test it is deemed fault free. Otherwise, based on the response of the first test set, a few additional test vectors are computed on the fly and applied to the circuit. These additional vectors are supposed to diagnose the fault in question. Such adaptive testing is also known as two-step diagnosis (Cheng et al., 1990).

Two-step diagnosis seems attractive at first, but has some severe problems. First of all, compared to single-step strategies, where a single test set is applied, adaptive T&D is slower. The reason is the online computation of additional diagnosis stimuli. If the computation is performed by software, it is easy to understand why it will be slower. Another solution is having specific hardware to carry out these calculations. Configurability is a desirable feature for this hardware, because test sets vary for different methods and circuits. Approaches based on FPGAs are a possibility, but such methods are still unexplored.

There is a more subtle but nonetheless powerful argument against adaptive diagnosis (Parker, 1992). The fault models on which test generation methods are based always assume that the behavior of faults is time invariant. For real faults this is not necessarily the case. Shorted nets often produce intermediate voltage levels, which are interpreted differently by the involved receivers, under the influence of several time dependent factors such as temperature and noise. The presence of hysteresis is also very common in faulty electronic circuits. Since an adaptive T&D method responds in accord to the fault behavior, if the behavior of the fault is time variant, so will the behavior of the adaptive method. Working with a time variant test set becomes considerably more difficult. If it is not possible to diagnose a fault the first time the scheme is applied, the diagnostician must bear in mind that, due to a possible time variant fault behavior, the generated diagnosis vectors may differ when the scheme is run again. Hence, the initial intent of speeding up diagnosis with an adaptive algorithm can be frustrated by the extra time spent in fault isolation.

In face of the difficulties of adaptive T&D, this work focuses solely on *one-step diagnosis*: all test vectors are applied and the results are analyzed only after all responses are collected. If there is a time variant fault, only the responses may change when a scheme is run more than once; the applied test vectors are fixed.

1.6.6 Testing digital clusters

In a completely digital EA, BS has the potential of providing an adequate T&D solution. Nevertheless, this is only possible if the EA is entirely composed of BS compliant ICs. The real scenario is often different. Common EAs also contain BS non-compliant digital, analog and mixed-signal ICs, and a great deal of discrete components. How can EAs be diagnosed in these circumstances?

Areas of BS non-compliant digital circuitry are called *digital clusters* or *glue logic*. Diagnosis in digital clusters becomes indistinguishable from ordinary diagnosis of digital circuits, which is a very hard problem (Abramovici et al., 1990; Rogel-Favila, 1991). Specific approaches for diagnosis of interconnects in digital clusters have been proposed in the literature (Barnfield and Moore, 1989; Hassan et al., 1989; Melton and Brglez, 1992). Compared to simple wire interconnect test sets, cluster tests can be considerably bigger in size. Digital cluster test sets are applied using the BSCs and the EA's primary I/Os. If the digital clusters are complex, the size of their test sets can easily become a test time bottleneck. This problem is likely to persist while there are chips where the inclusion of BS can lead to an unacceptable cost. Memory chips are a good example of this (de Jong and van Wijngaarden, 1992).

Nonetheless, the general tendency is that the use of glue logic declines for various reasons: (1) full BS design requirement (2) rapid development of interface protocols, and (3) use of reconfigurable computing technology. In regard to reason (1), manufacturers are beginning to require testing methodologies calling for BS and BIST in each component (Basset et al., 1990). In regard to reason (2), an example is the use of the peripheral component interconnect (PCI) standard; this standard has greatly reduced the need for glue logic interfaces (Michel, 1994). As for reason (3) just note that FPGA technology is now approaching the ten million gate milestone. With such technology, complex reconfigurable systems can be implemented on large FPGAs, whereas smaller FPGAs may be used to implement programmable interfaces between application specific integrated circuits (ASICs). According to this scenario, no glue logic components will be needed. Since the FPGAs themselves are normally BS compliant, full BS testable assemblies will be obtained. Additionally, if some ASICs in these EAs are upgraded with newer versions, it is likely that the FPGA interfaces can simply be reprogrammed with no need to change the rest of the hardware. It is also possible that ASIC cores will in the near future be surrounded by a fringe of programmable devices, intentionally put there to facilitate the interface with the external world, and to reduce the need for glue logic.

1.7 Other standards of the 1149 family

1.7.1 IEEE 1149.4

So far the existence of analog and mixed analog/digital components in the EAs has been completely ignored. However, the real world is mostly an analog world, and EAs containing analog and mixed analog/digital ICs are as common as those containing purely digital ICs. In fact, analog and mixed-signal T&D is currently an active research and development area (Vinnakota, 1998).

The T&D problem for *analog and mixed-signal clusters* is not addressed by the IEEE 1149.1 standard. Analog circuits require analog T&D signals, not available with BS (Lee, 1993; Matos et al., 1993; Parker et al., 1993; Parker, 1998; Thatcher and Tulloss, 1993). For this purpose, 1149.1 has been complemented with another standard of the same family — the IEEE 1149.4 Standard Mixed-Signal Test Bus (IEEE, 1999; Osseiran, 1999; Vinnakota, 1998). This recent standard, approved in 1999, provides analog test access to the I/Os of analog and mixed-signal chips. With 1149.4, analog and mixed-signal EAs can be tested in a similar way to digital BSEAs.

The architecture of the IEEE 1149.4 mixed-signal test bus is shown in Figure 1.4. Apart from the analog test signals AT1 and AT2, and the Test Bus Interface Circuit (TBIC), this architecture looks similar to the IEEE 1149.1 architecture. In fact, it is compatible with 1149.1, and enhances it with analog test facilities. Note that, analog wire interconnects between 1149.4 compliant ICs can be tested and diagnosed using solely the 1141.1 features. Only when analog components are tested will the 1149.4 features be necessary.

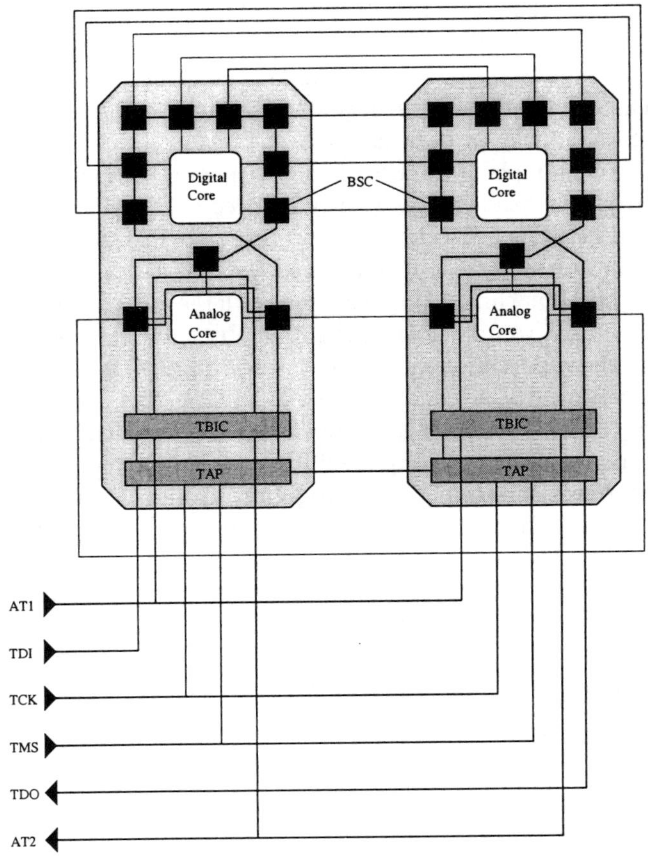

Figure 1.4. The mixed-signal test bus architecture.

The off-chip bus AT1 is connected in parallel to all compliant ICs in the EA. AT1 is used to convey a test current or voltage to a pin of an analog IC. AT2 is used to probe and measure a current or a voltage in the EA. The signals AT1 and AT2 interface with the ICs through the TBIC and, although not explicitly shown in Figure 1.4, are connected to the on-chip signals AB1 and AB2, respectively. The BS digital infrastructure is used to select the pin on which the test signal AB1 is applied and the pin on which the measurement is collected and sent back to the tester via AB2. The TBIC's main function is to isolate the chip under test, but it is also used for buffering purposes.

This is, very briefly, how IEEE 1149.4 operates. The standard can be used to (1) test and diagnose wire interconnects between analog pins, (2) test and diagnose active or passive discrete analog components in the EA, and (3) test and diagnose the mixed-signal ICs in the EA. Applications (1) and (2) are indispensable in post-manufacture T&D. Application (2) is considerably more expensive than application (1), since it is the BS equivalent of parametric testing. As in the 1149.1 standard, (3) is very expensive and should be restricted

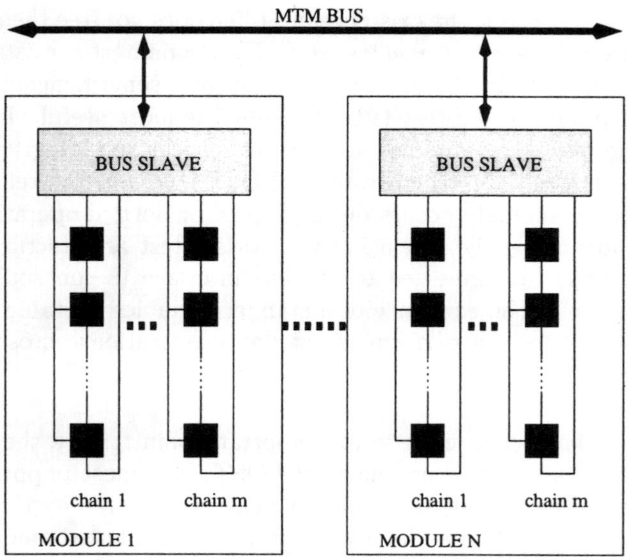

Figure 1.5. The module test and maintenance bus architecture.

to maintenance purposes only. For post manufacture test, one should either rely on the quality of the ICs or perform incoming inspection. According to the evolutionary trend, it will soon be possible to integrate many more types and higher quantities of analog devices in the same die. In this light, the T&D of discrete analog components will see its importance diminished, while the T&D of simple wire interconnects will gain even more relevance.

1.7.2 IEEE 1149.5

System test has been mentioned a few times. It is now time to examine a systematic scheme for testing and diagnosing a full system, composed of several BSEAs or modules. While IC level and assembly level T&D are oriented towards post manufacture test, system level test is oriented more towards maintenance and field service.

The IEEE 1149.5 Standard Module Test and Maintenance Bus (IEEE, 1995) specifies a normalized solution for system T&D. The main idea is to capitalize on boundary-scan assets and build a system test hierarchy (Jarwala et al., 1992). The general architecture can be seen in Figure 1.5. Various modules can be addressed individually or in groups, so that test stimuli are passed from a bus master (not shown in the figure) to the respective bus slaves. Bus slaves are implemented by an interfacing IC. Random module access permits more flexibility and speed than simply connecting the BS chains in each module to form a system level chain. Note that it is common to have optional cards in a system, in which case the respective module slot may be left vacant.

In post-assembly test, the existence of 1149.5 does not free the manufacturer from the conventional *functional* test step. Functional test consists of switching the system on, running the operating system and some benchmark software. However, in power-on self-test (POST) 1149.5 is most useful. POST quickly verifies if the necessary modules are present, checks the integrity of their BS chains, and performs *interconnect tests.* POST increases the reliability of the system. If some anomaly occurs during POST or normal operation, more accurate diagnostics can be run using the system test architecture. In fact, a proper tester can be connected to the bus interface to run sophisticated diagnosis programs. The existence of a standard enables portable testers to be used. These testers can perform maintenance operations across a variety of different systems.

The above discussion leads to an important point: since the interconnect test set is a fundamental component of POST, it is most important if it can be kept small. The module tests may be generated/stored by the bus master (Jarwala and Yau, 1991; Landis et al., 1992) and distributed among the slaves. Alternatively, they may be managed by the slaves themselves in a decentralized fashion (Su and Jou, 1999). In either case, having a small test set can simplify the hardware and reduce the POST application time.

1.8 The problem of interconnect faults

As seen in the previous section, there are still many difficult problems to be solved in boundary-scan T&D. Some of them are, perhaps, more pressing than the WITD problem. It may seem at a first glance that, because of the full controllability/observability provided by BS, the WIDT problem is trivial compared to other current problems. Effective solutions for the WIDT problem are already commercially available, whereas solutions for the problems posed by non BS components are still being investigated. However, as non-compliant components tend to disappear, the WITD problem will become the main problem of BS testing. Then, the techniques discussed in this work will make a tremendous difference in the accuracy and efficiency of EA test and diagnosis.

As the complexity of electronic circuits and systems increases, the use of pre-existent intellectual property (IP) core building blocks becomes more common. It is likely that all these building blocks incorporate BS. It is also likely that these components are supplied with their own test subprograms or self-test facilities. In such a vision, the role of the test engineer will be that of assembling the component test subprograms into a single test program to be applied through BS. However, in doing so one must develop tests for the wire interconnects between the blocks. Then, according to the authors' view, as systems get more complex, the goal will be to minimize the interconnect test data and maximize the diagnostic resolution of the test methods.

1.8.1 Brief background

In BS, the amount of test data depends obviously on the size of the CUT, and on the number of test vectors. This number corresponds to the length in bits of the serial test vectors that are applied to each net. Hereafter, the designation of *serial test vector length* and the symbol p will be used for this quantity. Minimizing test data means minimizing p, and maximizing DR means minimizing the ambiguity in diagnostic information.

In general terms, the traditional approach to the problem of generating BS test sets for interconnect diagnosis is the following. It is assumed that (1) any net can be affected by an open fault and that (2) any nets can be involved in a short fault. The former is generally true. The latter is a pessimistic assumption: shorts can only involve physically close nets. Nevertheless, traditional BS test sets are generated using these assumptions, together with simple *fault models* that describe the behavior of the faults.

These traditional schemes have been called *behavioral methods.* Several techniques for deriving behavioral schemes for diagnosis of interconnects will be discussed in detail in Chapter 3. Not considering which shorts are realistic makes these methods layout independent, which can be an advantage if handling layout information is problematic.

There are, however, two problems associated with behavioral methods. First, the optimization of the test vector length p is limited by the assumption that shorts can involve any nets. Second, the fact that the diagnostic information is solely the identification of faulty nets can make locating the actual defects or failures quite ambiguous. The result is a slow down of the process of isolating the defects and failures to be repaired.

Alternative methods that take into account which shorts are physically possible have emerged since the 90's. As opposed to behavioral methods, these schemes have been called *structural methods.* The introduction of structural information enables further optimization of diagnosis schemes. Additionally, because the locations of the realistic shorts have to be extracted beforehand, these data can be used later on in the form of high quality diagnostic information. This contributes to enhance DR.

Garey et al. were the first to propose a structural approach for optimizing short detection tests in bare PCBs (Garey et al., 1976). Their approach is based on *graph coloring algorithms.* Fourteen years later, Cheng et al. unearthed this technique, and suggested its extension to diagnosis rather than just detection (Cheng et al., 1990). However, they did not completely solve the problem, and improvements based on *graph color mixing* have been proposed since then by other researchers. These will be thoroughly discussed in Chapter 4.

The idea of using structural information for test purposes has also been proposed in the IC domain, where it is known as inductive fault analysis (IFA). There are two IFA flavors: the random approach (Ferguson and Shen, 1988; Shen et al., 1985) and the deterministic approach (Sousa et al., 1991; Teixeira et al., 1990). The random approach is a kind of Monte Carlo simulation where defects with a random size are randomly scattered over the layout. This approach makes sense for ICs where defects of random size and location can occur in the layout. It also gives some measure of the occurrence probability of defects. However, large numbers of defects have to be generated in order to study an adequate number of possible faulty situations and their probabilities. This limits the size of the circuits that can be analyzed.

On the other hand, the deterministic approach consists of scanning the layout to evaluate locations where defects can occur. In this way, the set of realistic defects can be extracted efficiently. Also, while doing this, the fault occurrence probabilities can also be extracted by means of computing fault *critical areas*. The critical area of a fault is the layout area in which the landing of a defect of average size can cause a fault (de Gyvez, 1991; Gandemer et al., 1988; Ferris-Prabhu, 1985).

In PCBs, where solder defects are dominant (Johnson, 1993), a deterministic IFA approach can be used very efficiently. Note that solder defects can cause shorts and opens only in the vicinity of solder points. Thus, one open defect is extracted for each solder joint, and short defects are extracted by evaluating the proximity of neighbor nets.

1.8.2 The contributions of this research

This work gives several original contributions to the field of test and diagnosis of wire interconnect faults in electronic assemblies. A summary of the main contributions is provided next.

- Fault models for interconnect nets are revised and upgraded. Multiple shorts, opens, or combinations of the two are considered. The models are classified as *analytic* or *synthetic* according to their use in response analysis or stimuli synthesis, respectively. Analytic models can be very general, whereas synthetic models need to be more constrained to simplify the generation of test vectors. The traditional fault models used in diagnostic synthesis are upgraded, so they can be used in structural diagnosis.

- A critical and formal analysis of behavioral diagnostic schemes for interconnect diagnosis is presented to support the more general approach to fault modeling. A diagnostic scheme may have different degrees of diagnostic resolution, but must be at least capable of detecting any non-redundant fault, single or multiple. The minimum requirement for test completeness (detection of any fault of any multiplicity) is identified, and the diagnostic properties of the most important diagnostic schemes are analyzed for the fault models considered. A new scheme, the *N sequence*, has been derived. This scheme is the shortest *diagonal independent* scheme. [2]

- The few existing structural methods for interconnect diagnosis are analyzed in the light of the new fault modeling approach. Problems in the schemes

[2]The diagonal independence property will be explained in Chapter 3.

based on graph coloring and graph color mixing are identified. To address these problems, a solution based on *statistical* graph color mixing is proposed.

- A theory of diagnostic resolution assessment based on statistical evaluation of fault diagnosability is introduced. An algorithm for interconnect diagnosis simulation derived from this theory is described. The possibility of an integrated automatic repair process is discussed, enabled by the high diagnostic resolution offered by structural methods.

- Tools that implement the algorithms originated in this work have been developed and experimental results derived. These tools use a deterministic algorithm for extracting realistic faults from PCB layouts and a statistical diagnosis simulator. The results clearly illustrate the capabilities of various schemes, and demonstrate the superiority of the proposed schemes.

In the proposed schemes two main limitations have been identified: (1) the sensitivity to the *personality of the circuit*, and (2) the lack of efficiency of some algorithms for large examples. As for (1), the capability of optimizing the test vector length p in the proposed structural scheme can be affected if the circuit contains nets other than ground or power nets that can short to an excessive number of other nets. Examples of such nets are long buses or nets connecting various components in parallel. Nonetheless, this effect has not been observed in the available benchmark circuits. [3] In regard to (2), the difficulty in finding benchmarks for validating the proposed algorithms provided little motivation for improving their efficiency. Nevertheless, since most algorithms work on graph representations of the interconnect circuit and its faults, randomly generated graphs have been used to represent hypothetically large circuits. In these artificially generated examples the inefficiency of the implemented algorithms has been exposed. However, improved versions of the algorithms are described, which would cope with PCBs of any practical size. The order of complexity for the improved versions has been computed to support this claim.

1.9 Summary

In this chapter the background material for understanding the problem addressed in this book has been provided and the main contributions of this research have been outlined.

Electronic assemblies and the inevitability of production defects have been discussed. The concept of production yield is explained and the principal types of faults affecting EAs are enumerated. From these, interconnect faults are the most likely. Test and diagnosis has been described, and the importance of optimizing test data size and diagnostic resolution has been emphasized.

[3] Ground and power nets were assumed to be signal nets just for studying this problem. Because these nets have a high fanout this problem was observed very clearly. However, very high fanout signal nets have not been found in the benchmarks used.

Some possibilities for performing automatic repair of EAs have been explored. It is concluded that complete and high resolution diagnostic information are key factors in automatic repair technology, and have a major impact on reducing the cost of automatic repair.

In-circuit testing and the role it has had in testing and diagnosing EAs has been discussed. It is explained that the increasing complexity and miniaturization of EAs is causing ICT to be replaced by boundary-scan testing.

The longest section in this chapter is dedicated to introducing boundary-scan testing, as defined by the IEEE 1149.1 standard. This is understandingly so, since this book is about testing and diagnosing interconnects in a BS environment. A readership already familiar with BS is targeted, but the material discussed does not really require a detailed knowledge of the standard. For those not familiar with BS, this introduction should more than suffice to enable understanding of the rest of the book. The architecture of BS is described and its operation explained. Then, many important BS related issues are discussed: test data serialization, the problem of applying power to shorts, adaptive T&D algorithms and glue logic testing.

Other standards related to BS are also addressed. Because EAs can contain analog and mixed-signal components, some consideration is given to standard IEEE 1149.4. This standard regulates T&D of analog and mixed-signal EAs. It is explained how 1149.4 is perfectly compatible with 1149.1, which is particularly important for wire interconnects: their testing and diagnosis can be carried out in the same way with both standards. Additionally, brief consideration is given to standard IEEE 1149.5, which refers to system T&D. The importance of compact interconnect test sets used during power-on self-tests has been stressed.

The central theme of this book is the optimization of test data size and/or diagnostic resolution of wire interconnect faults in BS assemblies. This chapter highlights the relevance of this problem, both as seen today and in the near future. A retrospective of the work done on this subject is briefly given before these matters are treated in detail in the subsequent chapters. Also, the contributions of this research to the solution and understanding of this problem are summarized. The innovations on fault modeling, behavioral diagnosis, structural diagnosis and diagnostic resolution assessment are anticipated. The structural scheme and the diagnosis simulation method introduced by this research are specially emphasized. For an outline of the organization of the book the reader is referred to the last section of the preface.

2 INTERCONNECT CIRCUIT AND FAULT MODELS

In this chapter models for the interconnect circuit and interconnect faults in electronic assemblies (EAs) are presented. It is assumed that boundary-scan (BS) is being used, so that any net can be driven or sensed using the boundary-scan cells (BSCs) or the EA's primary I/Os. Using conventional terms from the discipline of electronic testing, each net is fully *controllable* and fully *observable*. Models for short and open faults of any multiplicity are considered, including mixed short-open faults. The problem of fault masking is also addressed.

Fault models have been placed in two major categories: *synthetic* and *analytic* models. The synthetic fault models are used to generate diagnostic stimuli and are the models traditionally used: the *stuck-at-logic*, *wired-logic*, and the *strong-driver-logic* model. The strong-driver-logic model is not very commonly used. However, as will be shown in Chapter 4, this fault model is indispensable in structural interconnect diagnosis. The analytic fault models serve to analyze test responses and represent an original contribution of this research. Two analytic fault models are proposed: the *same-response* and the *any-response* models. The former is used to analyze shorts, whereas the latter is used for analyzing open faults.

In this chapter one begins to make use of some of the numerous symbols that constitute the notation of this book. These symbols will be introduced as needed, but, should the reader forget the meaning of a particular symbol, a complete list is given in the last pages of the book for a quick reference. The chapter has the following structure: Section 2.1 presents the interconnect circuit model and Section 2.2 presents the interconnect fault models. A summary is given in Section 2.3.

2.1 Interconnect circuit model

According to the assumptions given in Section 1.6, the EAs under consideration only contain BS compliant ICs interconnected by wires. Each interconnect net will be driven by an output BSC or an EA primary input, and the applied signal will be received by an input BSC or an EA primary output. The *interconnect circuit model* ignores ICs and considers the interconnects by themselves.

Figure 2.1 shows an 8-net interconnect circuit, where all net drivers are placed on the left hand side and identified with the letter "D" and all net receivers are placed on the right hand side and marked "R". The interconnects are labeled n_1 to n_8, the power net is labeled n_P and the ground net is labeled n_G. Shorts of other interconnects and these nets are treated differently from shorts involving only signal nets. Alternatively, n_P and n_G could be treated as pseudo signal interconnects driven by a constant signal, and thereby producing a constant response (0 for ground nets and 1 for power nets).

In Figure 2.1, any net is driven by a single driver and sensed by a single receiver. In a real situation, a net can be driven by multiple drivers and sensed by multiple receivers like for example in a bus. The multiple drivers must be isolated from one another by means of tri-state buffers to avoid driving conflicts. Diagnosis of faults in this kind of nets must ensure that all net drivers are exercised and all net receivers are monitored (Her et al., 1992). A simple way of doing that, which will be assumed in this work, is to make every net driver repeat the test sequence derived for that net, while the others are at a high impedance state. In this way, the length of the initial test sequence is multiplied by the number of drivers, which provides additional motivation for minimizing the basic test vector length. Having assumed that, all drivers of a particular net will be represented by a single driver, and all receivers of a particular net will be represented by a single receiver. It is implicit that during response analysis it will be taken into account which of the drivers was active, and which of the receivers collected the response under analysis.

During BS testing, each *Parallel Test Vector* (PTV) is shifted in serially through the TDI pin, and then applied (in parallel of course) by the net drivers. From the point of view of a net driver, each new PTV brings a new logic level to the net. The bit sequence applied by a net driver is called a *Serial Test Vector* (STV). The bit sequence sensed by a net receiver is called a *Serial Response Vector* (SRV). If an interconnect circuit is fault-free, all SRVs will equal their respective STVs. In contrast, if the circuit is faulty, and the test input stimuli are able to detect the fault, one or more SRVs will differ from their respective STVs. Interconnect diagnosis is about deriving STVs and analyzing SRVs, in order to locate the defects in the circuit.

The set of all STVs applied to an interconnect circuit is organized in a matrix called matrix of test vectors (MTV) or simply *test set*. The set of all SRVs collected from an interconnect circuit is organized in a matrix called Matrix

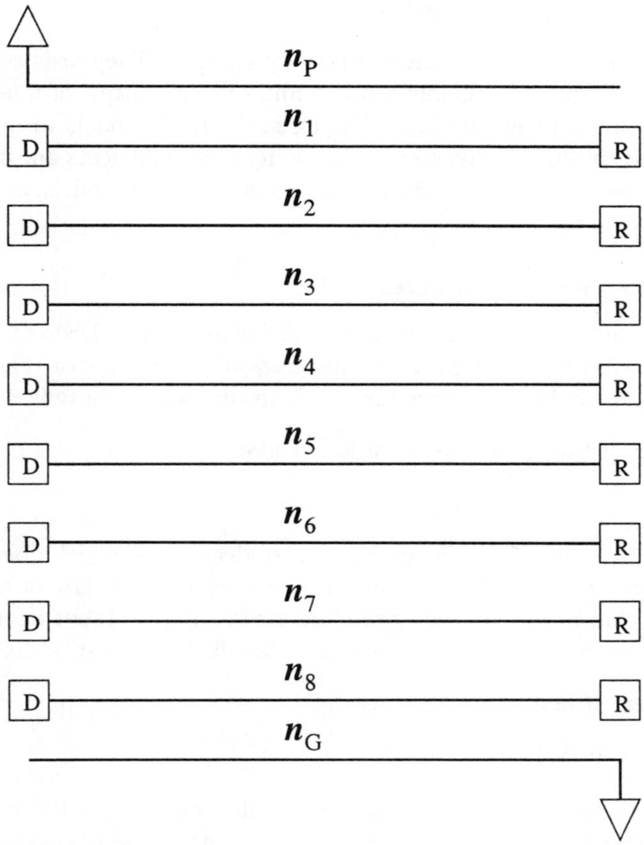

Figure 2.1. An 8-net interconnect circuit.

of Response Vectors (MRV) or simply *response set.* The following example illustrates these concepts.

Example 2.1 *For the 8-net circuit in Figure 2.1 a possible MTV is given by the following expression:*

$$MTV = \begin{bmatrix} 0 & 0 & 0 & 1 & 1 & 1 \\ 0 & 0 & 1 & 1 & 1 & 0 \\ 0 & 1 & 0 & 1 & 0 & 1 \\ 0 & 1 & 1 & 1 & 0 & 0 \\ 1 & 0 & 0 & 0 & 1 & 1 \\ 1 & 0 & 1 & 0 & 1 & 0 \\ 1 & 1 & 0 & 0 & 0 & 1 \\ 1 & 1 & 1 & 0 & 0 & 0 \end{bmatrix}$$

In this MTV, net n_1 is applied $STV_1 = 000111$, net n_2 is applied $STV_2 = 001110$ and so on up to net n_8, which is applied $STV_8 = 111000$.

2.2 Interconnect fault models

Fault models translate the behavior of faulty circuits. They are used to generate tests and diagnostic schemes, and to analyze the output of faulty circuits. Before studying interconnect fault diagnosis, the fault models to be used need be considered carefully. This section introduces faults, defects and failures, explains the difference between synthetic and analytic fault models, and describes the fault models themselves.

2.2.1 Faults, defects and failures

A *fault* is an alteration of the specified EA behavior, which violates its specification. A *defect* is a manufacture imperfection, which may or may not cause faults. Defects can be categorized in the following way (Wassink, 1989):

- *Cosmetic*: if the circuit behavior is not affected.
- *Operative*: if the circuit behavior is affected.

An example of a cosmetic defect may be a solder excess, which does not cause any short circuit or significant spurious impedance. A *failure* is a structural degradation that occurs during normal operation and is always operative. Operative defects and failures can be further classified in two sub-categories:

- *Redundant*: if faults are not caused.
- *Disruptive*: if faults are caused.

A redundant defect may be, for instance, a badly soldered joint that introduces a non negligible impedance, which, however, does not cause the circuit to violate its specification. Sometimes redundant defects may weaken the reliability of the EA, and later cause a fault. In terms of diagnosis, the concern is of course with disruptive defects and failures. These are of two basic types:

- *Short circuit*: when a spurious connection between two distinct nets is created.
- *Open circuit*: when the conductivity of a net is weakened [1]

A disruptive short circuit defect or failure between two nets n_i and n_j is represented by $e_{i,j}$. A disruptive open circuit defect or failure in a net n_i is represented by o_i. In Figure 2.2, some disruptive defects or failures are illustrated: o_3, o_5, o_5' and o_6 are opens; $e_{1,3}$, $e_{3,4}$, $e_{6,7}$ and $e_{7,8}$ are interconnect shorts; $e_{1,P}$ and $e_{8,G}$ are shorts to power and ground nets, respectively.

The distinction between defects and failures does not matter at this level, and one can simply talk about the faults they cause. For example, net n_5 in

[1] The name *open* suggests that a conductive path is actually broken, while this definition only requires the path to be weakened. This may be misleading, but it allows the conventional designation of *open fault* to be maintained while covering a broader range of behaviors.

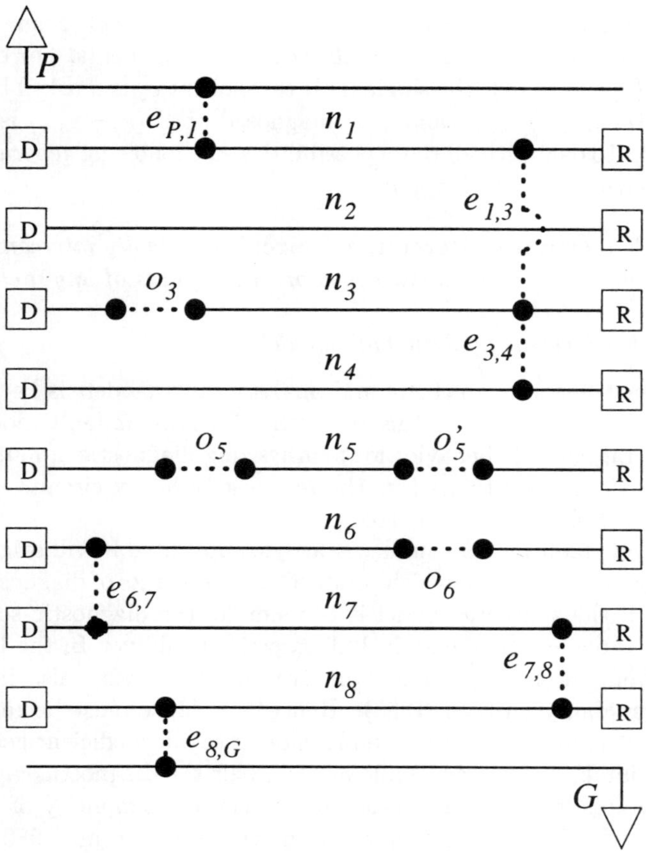

Figure 2.2. Open circuit and short circuit defects.

Figure 2.2 has two disruptive open defects or failures: o_5 and o_5'. Any one of them causes a *single open fault* in net n_5, which can simply be denoted o_5. Note that diagnostic procedures based on the observation of net receivers are unable to distinguish between o_5 and o_5'. The same applies for multiple short defects or failures between two nets n_i and n_j — they would cause a *single short fault* $e_{i,j}$. A *multiple fault* is a combination of single faults. Multiple faults can be classified in the following way:

- *Multiple short faults*: formed by a set of two or more single short faults.
- *Multiple open faults*: formed by a set of two or more single open faults.
- *Multiple mixed faults*: formed by a set of at least one single short and one single open fault.

Using the single faults from Figure 2.2, an example of a multiple short fault is $F_s = \{e_{3,4},\ e_{6,7},\ e_{7,8}\}$; an example of a multiple open fault is $F_o = \{o_5,\ o_6\}$;

an example of a multiple mixed fault is $F_m = \{e_{1,3},\ e_{6,7},\ o_5\}$. Some faults may have subcomponents that hide or *mask* other subcomponents. For example, in a fault $F = \{e_{1,3}, e_{3,4}, o_3\}$, the single fault component o_3 is masked by the fault component $\{e_{1,3}, e_{3,4}\}$ if o_3 cannot be diagnosed until $\{e_{1,3}, e_{3,4}\}$ is diagnosed and repaired. To deal with all these possibilities, the following general definition of *faulty interconnect circuit* applies:

Definition 2.1 (Faulty interconnect circuit) *A faulty interconnect circuit is a circuit which contains shorts, opens or mixed faults of any multiplicity.*

2.2.2 Synthetic versus analytic fault models

The distinction between synthetic and analytic fault models is key to the diagnosis methods developed in this research. A *synthetic* fault model is used to provide a target fault behavior for synthesizing diagnostic stimuli. An *analytic* fault model is used to analyze the response of faulty circuits, in order to determine which faults have occurred.

If a synthetic fault model is unnecessarily complicated it will originate cumbersome diagnostic schemes, which produce large amounts of diagnostic stimuli. In contrast, if the synthetic model is too simple, the diagnostic schemes will be more tractable but they may lack diagnostic capability. In the IC domain, the use of fault models of different nature and complexity also impacts the diagnostic capability (Aitken, 1995). Hence, a balance must be reached: the synthetic fault models should be simple enough to allow efficient generation of diagnostic stimuli, but not as simple as to impair the diagnostic capability. In the electronic testing field, the trade-off between the simplicity of fault models and their effectiveness has long been known (Galiay et al., 1980; Wadsack, 1978). Fortunately, it is also known that tests based on simple fault models can often detect faults having more complex behaviors (Huisman, 1995).

The analytic fault models arise from the observation that when analyzing the response of a faulty circuit some constraints of the synthetic fault models must be relaxed in order to diagnose the faults. The well defined synthetic fault models, although useful for simplifying the generation of diagnostic stimuli, lack enough flexibility for analyzing responses from actual faults. Thus, switching from synthetic to analytic fault models during response analysis is a means to effectively deal with the trade-off between the simplicity and capability of the fault models.

2.2.3 Short fault models

A short circuit defect or failure causes two distinct nets to be connected by a spurious resistance path. If the defect or failure resistance is low enough, then a short fault may occur. [2] Before introducing the synthetic and analytic fault models for shorts, the concept of *set of shorted nets* is defined:

---

[2]It is also possible that the short resistance is high and causes the fault to be redundant.

Definition 2.2 (Set of shorted nets) *A set of shorted nets S is a set of nets electrically connected by short faults, which respond with identical sequences when a test set is applied.*

The short response SRV_S is a function of the test vectors STV_i of the nets $n_i \in S_g$, where S_g is the set of drivers in S that generate the response SRV_S. There may be nets involved in a short whose drivers do not contribute to SRV_S because they are masked by open faults. Conversely, the response SRV_S may not necessarily be observed from every net in the short; there may be open faults preventing SRV_S from being observed from some receivers.

Synthetic short models. The synthetic short models are based on the traditional fault models for interconnect diagnosis: the stuck-at-logic, the wired-logic and the strong-driver-logic models. To account for masking effects, it is considered that these behaviors may be caused by a subset $S_g \in S$, rather than by all nets in S. The synthetic fault models are introduced by the following definitions:

Definition 2.3 (Stuck-at-logic fault model (S-A-0/1)) *According to this model, if at least one net in a set of shorted nets S is shorted to ground or power, S will respond with the vector $SRV_S = \underline{0}$ (S-A-0) or $SRV_S = \underline{1}$ (S-A-1), respectively.* [3]

Definition 2.4 (Wired-logic fault model (W-A/O)) *In this model, a set of shorted nets S will respond with $SRV_S[t] = \prod_{n_i \in S_g} STV_i[t]$, $t = 1, 2, ..., p$ for W-A behavior or $SRV_S[t] = \sum_{n_i \in S_g} STV_i[t]$, $t = 1, 2, ..., p$ for W-O behavior.* [4]

In practice, the S-A-0/1 models are good models for shorts to ground and power nets. The W-A/O model is historically associated with earlier technologies such as the Transistor-Transistor Logic (TTL) or Emitter Coupled Logic (ECL). In these technologies these models are quite accurate because of the asymmetry in the driving forces of the two logic levels 0 and 1. With the more recent Complementary Metal-Oxide-Semiconductor (CMOS) technology, the W-A/O model is less accurate (Galiay et al., 1980; Wadsack, 1978). In fact, for the CMOS technology the driving force for logic 0 can be comparable to the driving force for logic 1, and the resulting behavior cannot be captured by a simple model such as the W-A/O model. Fortunately, tests derived for the W-A/O model can still detect and diagnose faults having more complex behaviors. For this reason, the W-A/O and S-A-0/1 models are still the most

[3]The notation $\underline{0}$ represents a vector of zeros, i.e., if $SRV_S = \underline{0}$ then for $SRV_S[t] = 0$, $t = 1, 2, ..., p$, where t is the discrete time during test application, and p is the number of applied PTVs. In the same way, $SRV_S = \underline{1}$ is a vector of ones: $SRV_S[t] = 1$, $t = 1, ..., p$.

[4]W-A stands for Wired-And, while W-O stands for Wired-Or. Accordingly, $\prod$ represents the logical product AND, and $\sum$ represents the logical summation OR.

widely accepted models in the industry. However, as will be shown in Chapter 4, with the new structural diagnosis techniques these fault models become inadequate. In structural diagnosis, the W-A/O and S-A-0/1 models need to be complemented with the *strong-driver-logic* model (Park, 1996; Sousa et al., 1996b), which is defined as follows:

Definition 2.5 (Strong-Driver-logic fault model (S-D-k)) *According to this model, a set of shorted nets S will respond with the vector $SRV_S = STV_k$, where $n_k \in S$ is called the* strong-driver *net.*

The complexity of diagnosis either does not increase or increases only slightly with the addition of the S-D-k model. In return, the resulting test sets have a considerably better diagnostic capability.

Instead of considering S-A-0/1 models for shorts to ground and power nets, these faults could be modeled by strong-driver-ground (S-D-G) and strong-driver-power (S-D-P) models. However, since ground and power nets are not treated as signal interconnects these faults are instead modeled with S-A-0/1 models. One may argue that using the S-D-G and S-D-P models would have been a more elegant solution. Nonetheless, the two are equivalent, and the choice of using the S-A-0/1 model has prevailed for historical reasons.

Shorts to clock networks are not considered. First of all, these shorts usually cause catastrophic faults that are easy to diagnose by methods less sophisticated than BS. Delays in clock networks are often critical, which limits the usage of testability hardware associated with these nets. During BS interconnect test, only the BS specific clock signal TCK is active, and thus, this is the only clock nets can short to. However, shorts to TCK are easy to diagnose because they are likely to affect the integrity of the BS chain, which can be detected by BS integrity tests. Also, TCK is a strongly driven signal characterized by the distinctive 0-1 alternating pattern, which makes it easy to identify. For these reasons, shorts to clock networks are not taken into account. To close the discussion of synthetic short models, an illustrative example is presented:

Example 2.2 *Suppose $S = S_g = \{n_i, n_j\}$, $STV_i = 1001100$, and $STV_j = 0001000$. With the S-A-0/1 fault model $SRV_S = 0000000/1111111$. With the W-A/O fault model $SRV_S = 0001000/1001100$. Finally, with the S-D-i/j fault model $SRV_S = STV_i/STV_j = 1001100/0001000$.*

Analytic short models. To analyze the response of interconnect nets, the diagnostician reasons in the following way: *suspect nets that have the same response are potentially shorted.* The concept of *suspect net* will not be explained now because it depends on the specific type of analysis, behavioral (Section 3.1) or structural (Section 4.2). For the moment it should suffice to say that a suspect net is a net that may be faulty, but no mention will be made to how suspect nets are determined. Thus, the following definition of analytical short models is proposed:

Definition 2.6 (Same-Response fault model (S-R)) *In this model, any nets which respond with the same SRV are potentially shorted.*

The S-R analytic short model covers the behavior of any synthetic fault models, since all of them cause the SRVs of the shorted nets to be identical. For example, suppose that in a given process the behavior of shorts is determined by a majority function. That is, $SRV_S[t] = MAJ_{n_i \in S_g}(STV_i[t])$, $t = 1, 2, ..., p$. The majority function returns 1 if the number of 1s exceeds the number of 0s or returns 0 otherwise. Although all observable nets in the short will produce identical SRVs, it is unlikely that the response can be analyzed using the synthetic models described above. In many situations, the synthetic models would recognize the nets with erroneous responses as being faulty, but would naively fail to acknowledge that they might be shorted.

There is a special situation in diagnostic analysis where the S-R model should be complemented with the S-A-0/1 model: when a set of suspect nets responds with $\underline{0}$ or $\underline{1}$. In that case, the S-A-0/1 model should be applied to recognize that at least one net in S is potentially shorted to ground or power. The S-R model would recognize the potential set of shorted nets but would miss the possibility of short to ground or power. Moreover, the S-A-0/1 is the only model in the set that is used for analyzing shorts between single nets and ground or power nets.

For all its generality, the S-R model may still fail in a situation where the SRVs of the shorted nets disagree. Because of mismatches in input threshold levels of the receivers, it is actually possible that shorted nets respond with different SRVs. Two situations are possible: either no SRV is erroneous, and therefore the short will go undetected, or only some SRVs are erroneous, and the fault will only be partially diagnosed. Another situation where diagnostic analysis is difficult is when a non-deterministic fault occurs. These are faults that behave differently in different applications of the same test. The problems mentioned in this paragraph do not have specific solutions, to the best of the authors' knowledge. Hopefully, they will occur very seldom or, when they do, they will be diagnosed by tests derived for other faults.

2.2.4 Open fault models

An open circuit defect inserts a spurious resistance in a net's conducting path. If the resistance is high enough, then an open fault may occur. [5] The synthetic and the analytic open fault models are now introduced.

Synthetic open models. The synthetic fault model usually considered for open faults is the S-A-0/1 model adapted to open faults: if a net n_i is open, it responds with either $SRV_i = \underline{0}$ or $SRV_i = \underline{1}$. Other open behaviors may occur but they are difficult to predict and, since the S-A-0/1 is the most common behavior, they will not be considered in diagnostic synthesis. Besides, open faults are easy to expose with test vectors derived for other faults, and can easily be diagnosed with the analytic fault model discussed below. The S-A-

[5] It is also possible that the open resistance is low, and causes the defect to be redundant.

0/1 synthetic open fault model basically prevents the test vectors $\underline{0}$ or $\underline{1}$ from being used. Since the S-A-0/1 synthetic short fault model already does that, no synthetic open fault model is explicitly considered.

Analytic open models. The S-A-0/1 fault model is a good model for analyzing the response of open nets. Likely, open receivers will stabilize at a fixed level, which will be interpreted as a S-A-0/1 fault. The widespread use of pull-up and pull-down mechanisms for floating IC inputs also contributes to the effectiveness of the S-A-0/1 model. However, open faults may sometimes exhibit other behaviors, which justifies the proposal of the following analytic fault model:

Definition 2.7 (Any-Response fault model (A-R)) *Any suspect net is potentially open, regardless of the response it produces.*

Like for shorted nets, a net must be deemed a suspect net before it can qualify as a potential open net. Moreover, the suspicion of being open can accumulate with the suspicion of being shorted, in order to be able to deal with multiple mixed faults. Any ambiguities that may occur are resolved by the specific diagnostic analysis and fault isolation algorithms. Since they depend on the type of analysis, behavioral or structural, these algorithms will be discussed in Sections 3.1 and 4.2, respectively.

2.3 Summary

In this chapter interconnect circuit models and interconnect fault models have been presented. Throughout the remainder of the book these models shall be employed in describing the various interconnect diagnosis methods.

The interconnect circuit model assumes that BS is being used and consists of a set of nets having a single driver and a single receiver. Nets with multiple drivers and multiple receivers can appropriately be represented by nets with single drivers and single receivers. The diagnostic stimuli is applied by the driver and the responses are sensed by the receiver. The diagnostic stimuli (test vectors) and responses are synchronous digital signals, which can virtually be applied and received in parallel by the nets. [6]

A detailed analysis of multiple faults has been presented. Multiple faults are allowed to consist of exclusively shorts, exclusively opens or a combination of the two (mixed faults). Fault masking phenomena is also accounted for.

The interconnect fault models have been placed in two categories: synthetic and analytic models. The former are used when deriving diagnostic input stimuli and the latter when analyzing diagnostic responses. This original classification of fault models is then applied to the two basic fault types: shorts and opens.

---

[6]Note that the parallel application is only possible after the test vector is shifted in serially through the BS chain. The response vector is loaded in parallel into the BSCs, after which it is shifted out serially.

The synthetic fault models for shorts are the traditional wired-logic (W-A/O), the strong-driver-logic (S-D-k), and the stuck-at-logic (S-A-0/1) models. The S-D-k model is not commonly used in the industry. However, it will be shown in Chapter 4 that in structural diagnosis this model becomes very useful. Open faults only need the S-A-0/1 model as a synthetic fault model.

The following analytic fault models have been proposed: the same-response (S-R) and the any-response (A-R) models. The former allows interpreting any suspect nets that produce the same response as being shorted. The latter is utilized to identify any suspect nets as potentially open. The S-A-0/1 fault model is also used as an analytic model to diagnose shorts to ground or power nets.

3 BEHAVIORAL INTERCONNECT DIAGNOSIS

Diagnosis of interconnect faults comprises synthesis of test vectors and analysis of responses. In this chapter an original study of *behavioral* schemes for diagnosis of interconnect faults is presented. Behavioral diagnosis is performed with no knowledge of which faults are likely to occur. It is assumed that any net can be open, and that any two nets can be shorted. Behavioral schemes are very popular in the industry. They are simple, effective and can be derived independently from the layout design.

The study presented in this chapter draws from the more general fault models discussed in Chapter 2. Compared to previously published analyses, this study is more comprehensive as it takes into account mixed short-open faults of any multiplicity and deals with the problem of fault masking. Procedures for diagnostic analysis and fault isolation are studied. Then, several diagnostic synthesis schemes are analyzed in terms of test vector length and diagnostic capability. The theoretical properties of these schemes are formally proven by a set of lemmas and theorems, most of them being original. A new diagnostic scheme is presented: the *N sequence*. This sequence is the optimal sequence that possesses a property called the diagonal independence property, for both wired-AND (W-A) and wired-OR (W-O) short faults. The N sequence shows that W-A/O diagonal independence is an insufficient condition to ensure maximal diagnostic capability for the fault models considered.

This chapter is organized in the following way. In Section 3.1, a diagnostic analysis algorithm is examined. In Section 3.2, a fault isolation procedure is discussed. Section 3.3 presents a formal analysis of behavioral diagnosis synthesis. Section 3.4, presents a summary of the chapter.

3.1 Diagnostic analysis

The diagnostic analysis method presented in this section uses the analytic fault models of Section 2.2: the stuck-at-logic model (S-A-0/1), the same-response model (S-R) and the any-response fault model (A-R). In diagnostic analysis the matrix of response vectors *MRV* is analyzed and a diagnosis is issued. The information contained in the diagnosis should lead to a rapid isolation of faults to be repaired. A diagnostic scheme should at least be able to detect any fault. The concept of fault detectability is defined as follows:

Definition 3.1 (Detectability) *A fault F in a faulty EA is said to be detectable by a test set MTV if the response set MRV contains erroneous SRVs.*

Once it has been established that a scheme has detection capability, the scheme can be used in diagnosis. In the study of behavioral diagnosis presented in this chapter, the following concepts are needed:

Definition 3.2 (Set of suspect nets) *A set of suspect nets U_j is the set of all nets with an identical response SRV_{U_j}, which is erroneous for at least one net in U_j.*

Definition 3.3 (Partial diagnosis) *A partial diagnosis F'_j is the set of all single faults involving the nets of a set of suspect nets U_j. These are all single opens o_i of the nets $n_i \in U_j$, and all single shorts $e_{i,k}$ between every possible pair of nets $n_i, n_k \in U_j$.*

Definition 3.4 (Diagnosis) *A diagnosis F' for a fault F is the union set of all partial diagnoses F'_j of F.*

In a mature process a diagnosis F' will contain only a few partial diagnoses F'_j, most likely just one. Moreover, the number of nets in F' is also likely to be small. Based on these facts, a possible diagnostic analysis procedure is the one given by Algorithm 3.1 below. The algorithm applies the S-R, A-R and S-A-0/1 analytic fault models to each set of suspect nets U_j, in order to identify the partial diagnoses F'_j of F'.

The complexity of Algorithm 3.1 is determined by the complexity of identifying sets of suspect nets U_j (line 1). A simple way to identify the sets U_j is to compare each SRV with its corresponding STV. If the SRV and the STV differ, the net is included in a set of suspect nets U_j and the SRV is recorded to identify more nets with the same response. These nets are also included in U_j. Since SRVs and STVs are p-bit vectors, two such vectors can be compared in $O(p)$ time at most. To identify the sets U_j, all N nets in the circuit need be checked. Thus, the complexity of Algorithm 3.1 is $O(N \times p)$, since most likely there is only one set U_j. As will be seen, for the most commonly used diagnostic synthesis algorithms, p is in the order of $\log N$. Thus, the complexity of Algorithm 3.1 is $O(N \log N)$, which is adequate for the largest EAs (a few tens of thousands of nets). Some higher capability diagnostic synthesis methods produce test sets where p is in the order of N. In this case the complexity of Algorithm 3.1 becomes $O(N^2)$, which may be impractical for EAs as large as 10^4 nets.

Algorithm 3.1 (analyzeResponse(*MTV*, *MRV*)) :

$U =$ *identifySetsOfSuspectNets(*MRV*);*
$F' = \emptyset$; `/* initialize diagnosis */`
for each U_j **in** U **do** {
    $F'_j = \emptyset$; `/* initialize partial diagnosis */`
    if$(\#U_j == 1)$ `/* a single suspect net */`
        if$(SRV_{U_j} == \underline{0}/\underline{1})$ `/* apply S-A-0/1 and A-R */`
            $F'_j = \{e_{i,G/P},\ o_i : n_i \in U_j\}$;
        else `/* apply A-R only */`
            $F'_j = \{o_i : n_i \in U_j\}$;
    else `/* multiple suspect nets */`
        if$(SRV_{U_j} == \underline{0}/\underline{1})$ `/*apply S-R, S-A-0/1 and A-R*/`
            $F'_j = \{e_{i,k},\ e_{i,G/P},\ o_i : n_i,\ n_k \in U_j\}$;
        else `/* apply S-R and A-R */`
            $F'_j = \{e_{i,k},\ o_i : n_i,\ n_k \in U_j\}$; }
    $F' = F' \cup F'_j$;
return F';

A non-empty diagnosis F' obtained with Algorithm 3.1 for a detectable fault F can be classified as follows:

- *Complete* if $F' \supset F$.
- *Incomplete* if $F' \not\supset F \wedge F' \cap F \neq \emptyset$.
- *Incorrect* if $F' \cap F = \emptyset$.

Example 3.1 *Consider the 2-net interconnect circuit of Figure 3.1, which contains the mixed fault* $F = \{e_{1,2},\ o_1\}$*, and suppose* $SRV_1, SRV_2 \neq \underline{0}, \underline{1}$*. Next, three possible scenarios are considered, where three different diagnosis* F' *are obtained, each of them having different classifications.*

In the most likely scenario, $SRV_1 \neq STV_1$*,* $SRV_2 \neq STV_2$ *and* $SRV_1 \neq SRV_2$*. In this scenario Algorithm 3.1 produces the diagnosis* $F' = \{o_1, o_2\}$*, which is incomplete because open* o_1 *masks short* $e_{1,2}$*, preventing it from being diagnosed.*

In another scenario, the fault effect is such that, despite the presence of o_1*,* $SRV_1 = SRV_2$*. In this case,* o_1 *is also included in* F'*, which becomes a complete diagnosis* $F' = \{o_1, o_2, e_{1,2}\}$*.*

In the last scenario, some strange phenomenon causes $SRV_1 = STV_1$*,* $SRV_2 \neq SRV_1$ *and* $SRV_2 \neq STV_2$*. In this case an incorrect diagnosis* $F' = \{o_2\}$ *is produced.*

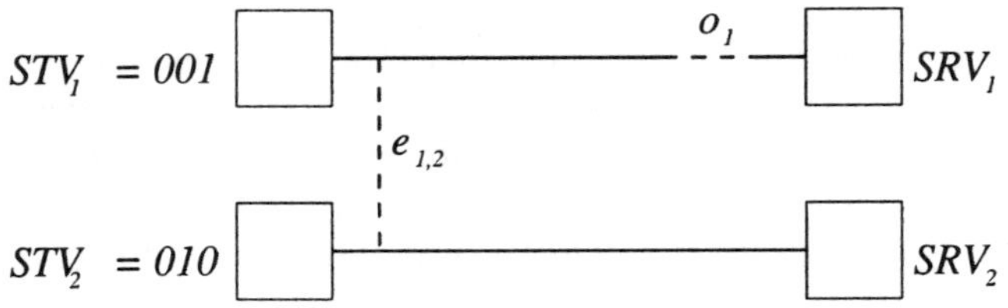

Figure 3.1. The effect of a multiple mixed fault.

Complete and incomplete diagnoses are further classified as follows:

- *Ambiguous* if $\exists f : f \in F' \wedge f \notin F$.
- *Unambiguous* if $\forall f : f \in F' \Rightarrow f \in F$.
- *Perfect* if a diagnosis is complete and unambiguous.

Diagnostic ambiguity is the inverse concept of diagnostic resolution and is directly related to fault isolation time. In behavioral diagnosis, diagnostic ambiguity increases rapidly with the size of the set of suspect nets $\#U_j$. According to Algorithm 3.1, two situations are possible: (1) when $\#U_j = 1$, the diagnostic ambiguity is low, since the only net $n_i \in U_j$ is either shorted to $n_{G/P}$ or open; (2) when $\#U_j > 1$, the shorts diagnostic ambiguity increases quadratically with $\#U_j$, [1] and the number of potential opens or shorts to $n_{G/P}$ increases identically with $\#U_j$.

3.2 Fault isolation

In fault isolation, the diagnosed faults are located in the EA using the information given in behavioral diagnostic analysis. When fault isolation is performed there is no guarantee that the diagnosed faults are actually present in the EA. The quality of the diagnostic information determines the effectiveness and efficiency of fault isolation, but the fault isolation procedure itself is also important. Algorithm 3.2 given below is a possible fault isolation procedure; it takes as input a diagnosis F', issued by Algorithm 3.1 for a circuit with a fault F, and tries to find the respective defects in the layout.

The most important aspect to notice about Algorithm 3.2 is the way fault masking is dealt with. Although F' may include potentially masked faults, the isolation procedure does not attempt to locate them initially. Only the *visible* faults are isolated and repaired, after which the EA is retested. In the most likely scenario, no more faults will be detected in the second time the EA is tested, and therefore no time is wasted trying to isolate non-existent masked faults. However, in some rare cases, masked faults do occur; the ones unmasked by the repaired faults are detected and diagnosed during the second time the

[1] The number of potential single shorts is given by $\binom{\#U_j}{2} = [(\#U_j)^2 - \#U_j)]/2$.

EA is tested; after they are repaired they may unmask other faults, and so on. The procedure iterates in this fashion until all masked faults are unmasked, isolated and repaired. For example, in line 10, Algorithm 3.2 is considering a partial diagnosis $F'_j = \{e_{i,k},\ o_i : n_i, n_k \in U_j\}$. Since all nets in U_j have an identical SRV, it is most likely that these nets are shorted. Thus, in line 11, the shorts $e_{i,k}$ among the nets of U_j are tentatively located. If found, they are marked for repair and the procedure continues with the next set of suspect nets U_j, without attempting to locate the opens o_i. If no shorts $e_{i,k}$ are found, the occurrence of the opens o_i is investigated in the next iteration in line 3. If not even the opens o_i can be located, the procedure reports failure, and the EA is either set aside for a more thorough inspection or discarded.

Algorithm 3.2 (isolateFaults(F')) :

1. **for each** `/*partial diagnosis*/` $F'_j \subset F'$ **do** {
2. **if**$(F'_j = \{o_i : n_i \in U_j\})$
3. **if**(*findDefects*(o_i) == *FOUND*);
4. *markForRepair*(o_i); **continue**;
5. **else return** *FAILURE*;
6. **else if**$(F'_j = \{e_{i,G/P},\ o_i : n_i \in U_j\})$
7. **if**(*findDefects*$(e_{i,G/P})$ == *FOUND*);
8. *markForRepair*$(e_{i,G/P})$; **continue**;
9. **else** $F'_j = F'_j - e_{i,G/P}$; **goto 2**;
10. **else if**$(F'_j = \{e_{i,k},\ o_i : n_i, n_k \in U_j\})$
11. **if**(*findDefects*$(e_{i,k})$ == *FOUND*);
12. *markForRepair*$(e_{i,k})$; **continue**;
13. **else** $F'_j = F'_j - e_{i,k}$; **goto 2**;
14. **else if**$(F' = \{e_{i,k},\ e_{i,G/P},\ o_i : n_i, n_k \in U_j\})$
15. **if**(*findDefects*$(e_{i,k})$ == *FOUND*);
16. *markForRepair*$(e_{i,k})$; **continue**;
17. **else** $F'_j = F'_j - e_{i,k}$; **goto 6**; }
18. **return** *SUCCESS*;

Algorithm 3.2 is roughly the procedure applied by a repair technician to locate defects using diagnostic information. The procedure is normally manually applied, as automation of fault isolation is still a difficult problem. Moreover, the technician is free to use common sense and does not need to follow the procedure exactly as outlined. For example, the technician may decide to mark other defects for repair based on information other than the response signals such as visual information, other measurements, etc.

It might be suggested that automatic repair (Section 1.4) can be accomplished by automating the execution of Algorithm 3.2. That would require a data structure representing the physical structure of the EA. Since in behavioral diagnosis no structural information is available, automatic repair is clearly a subject that fits better in structural diagnosis, and will be discussed in Chapter 4.

In behavioral diagnosis, the only way to improve diagnostic capability is by improving the test set MTV applied to the nets. As a matter of fact, different test sets may produce very different diagnoses F', which can match the fault F more or less accurately. Many behavioral schemes for generating diagnosis vectors for interconnects have been proposed in the literature. The most representative ones are discussed in detail in the next section.

3.3 Diagnostic synthesis

In diagnostic synthesis the goal is to simultaneously minimize diagnostic ambiguity and test vector length. In the literature, the analysis of diagnostic synthesis is usually restricted to W-A/O shorts. S-D-k shorts (Park, 1996) or multiple mixed faults (Lien and Breuer, 1991; Shi and Fuchs, 1995) are rarely considered. In the approach presented here, the more general fault models of Section 2.2 are considered. It is assumed that faults may have any multiplicity, and each fault subcomponent may behave according to the following synthetic fault models: the wired-logic (W-A/O), strong-driver-logic (S-D-k) and stuck-at-logic (S-A-0/1) models.

Since a fault must be detected before it can be diagnosed, some assurance should be given that the potential faults will be detectable. Naturally, it is not possible to do that without assuming certain fault behaviors. In other words, detectability can only be proven for a set of synthetic fault models. For our synthetic fault models, detectability is established by the following lemma and theorem:

Lemma 3.1 *In a test set MTV, where all STVs are unique and shorted nets behave according to the synthetic short models, a fault in which two or more nets have shorted receivers can always be detected.*

Proof: *The shorted receivers always produce identical SRVs, according to the synthetic short models. Thus, because the STVs are unique, at least one net n_i will have $SRV_i \neq STV_i$, which is sufficient for the fault to be detected.* ■

Theorem 3.1 *According to the synthetic fault models, the necessary and sufficient condition for detectability of any fault F is that all STVs in the test set are unique and different from $\underline{0}$ or $\underline{1}$.*

Proof: *To prove necessity, it should be noted that shorts between nets assigned identical STVs cannot be detected, and that S-A-0/1 faults cannot be detected for nets assigned $\underline{0}/\underline{1}$ vectors, respectively. To prove sufficiency, the technique of* reductio ad absurdum *will be used. Suppose there is a fault F which is undetectable. Then, there must be at least a faulty net n_i which either (a) has its driver disconnected from its receiver by an open fault or (b) is shorted to some other net n_j.*

In case (a) the receiver of n_i must be shorted to two or more net drivers to justify its correct response $SRV_i = STV_i$. Note that, according to the synthetic fault models and STV uniqueness, it takes at least two shorted nets to reproduce the response SRV_i, if that is at all possible. Let n_{i+1} be one of those nets. It must be true that the driver of n_{i+1} is disconnected from its receiver. Otherwise, *the receivers of n_i and n_{i+1} would be shorted and, according to Lemma 3.1, F would be detected. To justify the correct response $SRV_{i+1} = STV_{i+1}$ two or more net drivers need be shorted to the receiver of n_{i+1}. Assuming that n_i can be used, at least one more net is needed. If n_i cannot be used then at least two more nets are needed. Let n_{i+2} be one of those nets. As before, to avoid shorting the receivers of n_{i+1} and n_{i+2}, the driver of n_{i+2} must be disconnected from its receiver. Also, at least one more net n_{i+3} is needed to justify $SRV_{i+2} = STV_{i+2}$. This process continues ad infinitum: n_k always requires the addition of yet another net n_{k+1} to justify $SRV_k = STV_k$. Note that n_i can only be used once to justify some net n_k, $k > i$, and that the nets $n_{i+1}, n_{i+2}, \dots, n_{k-1}$ cannot be used because they are already shorted to a receiver. Therefore, an infinite number of nets is required for F to be undetectable, which is absurd.*

In case (b) it must be verified again that the receiver of the shorted net n_j is disconnected from its driver to avoid the short of receivers n_i and n_j. However, in that case, net n_j is in the same situation as net n_i in case (a). Note that n_i cannot be used to justify any other receivers, since it would short those receivers to its own. Thus, case (b) reduces to case (a). Then, since (a) and (b) are absurd, sufficiency is proven and it must be true that F is detectable.

■

Theorem 3.1 establishes that any fault F will be detected as long as it behaves according to the synthetic fault models. Another way of expressing this result is by stating that in interconnect circuits *cyclic masking relationships* (Abramovici et al., 1990) are impossible. To qualify as a behavioral diagnostic scheme detection must be guaranteed. Thus, the following definition of a diagnostic scheme is proposed:

Definition 3.5 (Behavioral interconnect diagnostic scheme) *A test set MTV is an interconnect diagnostic scheme if it verifies the minimum requirement for detectability imposed by Theorem 3.1.*

Once the detectability of any fault F has been ensured, the remaining question is whether F can be diagnosed, which is guaranteed by the following theorem:

Theorem 3.2 *According to the synthetic fault models, a fault F will be diagnosed in a finite number of test-diagnose-repair iterations if the following three conditions are satisfied: (1) the test set is a behavioral interconnect diagnostic scheme; (2) Algorithm 3.1 is used for diagnostic analysis; (3) Algorithm 3.2 is used for fault isolation.*

Proof: *To prove this theorem it is only necessary to prove that no diagnosis F' of F will be incorrect, i.e., $\exists f : f \in F \wedge f \in F'$. Then, because F has a finite number of single faults, it is guaranteed that after a finite number of test-diagnose-repair iterations, using Algorithm 3.1 for diagnostic analysis and Algorithm 3.2 for fault isolation, all single faults of F will be diagnosed.*

To prove that no diagnosis F' of F will be incorrect note the following. A consequence of Theorem 3.1 is that a detectable fault F has either unjustified opens or shorts between net receivers. According to Algorithm 3.1, those faults will appear in the final diagnosis F', which therefore cannot be incorrect. ■

Theorem 3.2 establishes the completeness of the whole diagnostic procedure for the synthetic fault models considered. Knowing that those faults will be diagnosed, the next question is how efficient diagnosis and fault isolation are. Diagnosis should identify the faults in a minimum number of test-diagnose-repair cycles, preferably in a single pass. A minimum time should be spent in each pass, which is determined by the attainable diagnostic resolution. The number of test-diagnose-repair iterations needed to identify and repair all faults in a faulty EA depends on the probability of fault masking, which is usually small; most of the times a single test step will be enough for interconnect diagnosis (Bleeker et al., 1993). Therefore, the efficiency of diagnosis and fault isolation is mainly determined by the diagnostic resolution of the method, which in turn depends on the choice of the interconnect diagnostic scheme.

Traditionally, diagnostic capability has been studied using the concepts of *aliasing* and *confounding* (Jarwala and Yau, 1989). Although aliasing and confounding do not provide a precise measure of diagnostic resolution, this framework has been very useful to analyze test sets. Aliasing and confounding are formally defined next, together with a definition of *faulty subset*, which is used in conjunction with confounding:

Definition 3.6 (Aliasing) *A diagnostic scheme is said to be aliasing with respect to some synthetic fault behavior if that behavior can generate a set of suspect nets U_j, which includes a fault free (aliased) net n_i.*

Definition 3.7 (Faulty subset) *A faulty subset V is a subset of a set of suspect nets U_j, whose response $SRV_V = SRV_{U_j}$ can be explained in terms of some synthetic fault model involving only the nets of V.*

Definition 3.8 (Confounding) *A diagnostic scheme is said to be confounding with respect to a pair of synthetic fault behaviors if those fault behaviors can create a set of suspect nets U_j, which can be explained by a pair of disjoint faulty subsets V_1/V_2 of U_j.*

According to Section 3.1, a diagnosis F' of a fault F is ambiguous if $\exists f : f \in F' \wedge f \notin F$. Thus, aliasing and confounding contribute to diagnostic ambiguity for the following reasons: since aliased nets are fault-free, any single fault f involving these nets will be such that $f \in F' \wedge f \notin F$; since confounding responses are produced by disjoint faulty subsets, then any single short f across these sets will be such that $f \in F' \wedge f \notin F$.

Aliasing and confounding are independent from fault masking. The fact that some ambiguities may be hidden by some masking fault does not mean they are not there. As soon as the masking fault is repaired, any aliasing or confounding hidden ambiguities will surface. It can also happen that fault masking introduces additional ambiguity: in Example 3.1 open o_1 masks short $e_{1,2}$ and introduces the non-existent open o_2 in diagnosis F', making it ambiguous. With or without fault masking, it is always important to minimize aliasing and confounding ambiguities.

In the subsections that follow, the most important behavioral test sets are discussed in terms of STV length and diagnostic capability, using the concepts of aliasing and confounding. From Section 3.3.1 through Section 3.3.7 the following schemes are analyzed: the modified counting sequence, the counting sequence plus complement, the walking sequences, the min-weight sequence, the self-diagnosis sequence, the global-diagnosis sequence and the N+1 sequence. In Section 3.3.8, the newly derived N sequence is presented.

3.3.1 Modified counting sequence

The modified counting sequence originates from the *counting sequence* test set (Kautz, 1974), which is simply given by $STV_i = (i-1)_2$. [2] Kautz simply had in mind the detection of shorts, and proposed that the STVs form a counting sequence from 0 to $N-1$. Hence the scheme's name, whose STV length is obviously given by

$$p = \lceil \log_2 N \rceil . \tag{3.1}$$

Example 3.2 *The counting sequence test set for an 8-net circuit is given by the following expression:*

$$MTV = \begin{bmatrix} 0 & 0 & 0 \\ 0 & 0 & 1 \\ 0 & 1 & 0 \\ 0 & 1 & 1 \\ 1 & 0 & 0 \\ 1 & 0 & 1 \\ 1 & 1 & 0 \\ 1 & 1 & 1 \end{bmatrix}$$

The fact that the test vector length is of complexity $O(\log N)$ means that the number of nets may increase exponentially while the test vector length increases only linearly. However, the counting sequence test set does not satisfy the minimum requirement for detectability as established by Theorem 3.1, and therefore cannot be considered a diagnostic scheme according to Definition 3.5. In fact, a S-A-0 fault in n_1 cannot be detected because $STV_1 = \underline{0}$, and a S-A-1 fault in n_N cannot be detected when $N = 2^p$ because $STV_N = \underline{1}$.

[2]The notation $(i-1)_2$ represents the natural number $i-1$ in the binary system.

In order to be able to detect S-A-0/1 faults, the STVs $\underline{0}/\underline{1}$ must be excluded from the counting sequence (Goel and McMahon, 1982). The test set obtained in that way was called the *modified counting sequence*, and its STV length is given by

$$p = \lceil \log_2(N+2) \rceil . \tag{3.2}$$

Example 3.3 *The modified counting sequence for an 8-net circuit is given by the following expression:*

$$MTV = \begin{bmatrix} 0 & 0 & 0 & 1 \\ 0 & 0 & 1 & 0 \\ 0 & 0 & 1 & 1 \\ 0 & 1 & 0 & 0 \\ 0 & 1 & 0 & 1 \\ 0 & 1 & 1 & 0 \\ 0 & 1 & 1 & 1 \\ 1 & 0 & 0 & 0 \end{bmatrix}$$

The STV length of the modified counting sequence differs little from that of the counting sequence, as can be concluded by comparing Equations (3.1) and (3.2). Because the modified counting sequence fulfills the minimum requirement for detectability imposed by Theorem 3.1, it qualifies as a diagnostic scheme according to Definition 3.5. Its diagnostic capability is as established by Theorem 3.3 below, which is proven using the next two lemmas:

Lemma 3.2 *Any diagnostic scheme MTV is aliasing-free for (a) S-A-0/1 and (b) S-D-k faults.*

Proof: *Clause (a) is proved by observing that any net n_i affected by a S-A-0/1 fault produces $SRV_i = \underline{0}/\underline{1}$, which is different from any other STV in MTV. Thus, S-A-0/1 faults cannot produce aliasing. For clause (b), note that the strong-driver net n_k cannot be aliased because it is faulty by definition, and that any other net n_i in the short cannot be aliased because $SRV_i = STV_k \neq STV_i$.*

■

Lemma 3.3 *If MTV is a diagnostic scheme, any pair of faulty subsets V_1/V_2 cannot confound if characterized by the behavior pairs (a) S-A-0/S-A-1, S-A-0/W-O, S-A-0/S-D-k, S-A-1/W-A, S-A-1/S-D-k and (b) S-D-k/S-D-l.*

Proof: *Clause (a) is proved by observing that $H(SRV_{V_1}) \neq H(SRV_{V_2})$.* [3] *Note that $H(SRV_{V_1}) = 0/p$ if V is a S-A-0/1 faulty subset, $H(SRV_V) > 0$ if V is a W-O faulty subset, $H(SRV_V) < p$ if V is a W-A faulty subset, and*

---

[3] $H(X)$ denotes the Hamming weight of bit vector X. The Hamming weight of a bit vector is the number of 1s in that vector.

$H(SRV_V) \neq 0, p$ if V is a S-D-k faulty subset. Clause (b) follows from $SRV_{V_1} = STV_k \neq SRV_{V_2} = STV_l$, which implies that V_1 and V_2 cannot confound. ■

Theorem 3.3 *The modified counting sequence test set MTV is (a) aliasing-free for S-A-0/1 and S-D-k faults and (b) confounding-free for the behavior pairs S-A-0/S-A-1, S-A-0/W-O, S-A-1/W-A, S-D-k/S-D-l, S-A-0/S-D-k and S-A-1/S-D-k.*

Proof: *Clause (a) follows from Lemma 3.2, and clause (b) follows from Lemma 3.3.* ■

Theorem 3.3 shows that, in the modified counting sequence, aliasing-freedom can be guaranteed only for S-A-0/1 faults, and confounding-freedom can be guaranteed only for the behavior pairs of Lemma 3.3. Due to the aliasing problem, the diagnostic capability of the modified counting sequence is indeed modest.

Example 3.4 *Referring to Example 3.3, suppose there is a set of shorted nets $S = \{n_4, n_5\}$, which behave according to the W-O behavior. S produces the set of suspect nets $U_j = \{n_4,\ n_5,\ n_6\}$ whose response is $SRV_{U_j} = 0111$. Net n_6 is fault free but is nonetheless a member of U_j. Therefore, according to Definition 3.6, the modified counting sequence is aliasing with respect to the W-O behavior. Moreover, n_6 is said to be an aliased net. It can also be shown that this diagnostic scheme is aliasing for the W-A behavior.*

3.3.2 Counting sequence plus complement

Wagner proposed an aliasing-free test set, which will be called here the *modified counting sequence plus complement* (Wagner, 1987). This test set is given by $MTV = [MTV^{MC}\ \overline{MTV^{MC}}]$, where MTV^{MC} is the modified counting sequence and $\overline{MTV^{MC}}$ is the matrix obtained by complementing each element in matrix MTV^{MC}. The test set is obtained by horizontally concatenating the two matrices MTV^{MC} and $\overline{MTV^{MC}}$, and the STV length is given by

$$p = 2\lceil \log_2(N+2) \rceil. \tag{3.3}$$

The Wagner test set was slightly optimized by Cheng et al. who noticed that the modified counting sequence test set MTV^{MC} can be replaced with the counting sequence test set MTV^{CS} when producing the test set MTV (Cheng et al., 1990). The resulting test set, given by $MTV = [MTV^{CS}\ \overline{MTV^{CS}}]$, may be called the *counting sequence plus complement* scheme. It achieves aliasing-freedom like the Wagner test set, but with a slightly shorter STV length:

$$p = 2\lceil \log_2 N \rceil. \tag{3.4}$$

Example 3.5 *For an 8-net circuit, the counting sequence plus complement test set is given by the following expression:*

$$MTV = \left[\begin{array}{ccc|ccc} 0 & 0 & 0 & 1 & 1 & 1 \\ 0 & 0 & 1 & 1 & 1 & 0 \\ 0 & 1 & 0 & 1 & 0 & 1 \\ 0 & 1 & 1 & 1 & 0 & 0 \\ 1 & 0 & 0 & 0 & 1 & 1 \\ 1 & 0 & 1 & 0 & 1 & 0 \\ 1 & 1 & 0 & 0 & 0 & 1 \\ 1 & 1 & 1 & 0 & 0 & 0 \end{array}\right]$$

The vertical dividing line illustrates how this test set is formed by concatenating the counting sequence matrix MTV^{CS} and its complement $\overline{MTV^{CS}}$.

The STV length of the Wagner test set is twice that of the counting sequence, and therefore still retains a logarithmic complexity $O(\log N)$. To establish the diagnostic capability of the counting sequence plus complement test set, the following definitions are needed:

Definition 3.9 (STV coverage) *STV_i is said to cover STV_j ($STV_i \supset STV_j$ or $STV_j \subset STV_i$) iff $\forall t: \ STV_j[t] = 1 \Rightarrow STV_i[t] = 1$.*

Definition 3.10 (STV independence) *Two vectors STV_i and STV_j are said to be independent iff $STV_i \not\supset STV_j \wedge STV_j \not\supset STV_i$.*

From the above definition it immediately follows that the STVs $\underline{0}$ and $\underline{1}$ cannot integrate a set of independent STVs. As a matter of fact, $\underline{0}$ would be covered by any other STV and $\underline{1}$ would cover any STV. Additionally, in a set of independent STVs all of them must be unique, as identical STVs would cover each other. Therefore, a test set of independent STVs is a diagnostic scheme according to Definition 3.5.

Definition 3.11 (Fixed weight diagnostic scheme) *A test set is said to be of fixed weight if $\forall i: H(STV_i) = W$, where W is a non-zero integer.*

Lemma 3.4 *A fixed weight diagnostic scheme MTV of weight W is a matrix of independent STVs.*

Proof: *With such an MTV, any two vectors STV_i and STV_j cannot have their 1 bits in the same positions or otherwise they would be identical, which is impossible for a diagnostic scheme. Thus $\exists t_1, t_2: \ STV_i[t_1] = 1 \wedge STV_j[t_1] = 0 \vee STV_i[t_2] = 0 \wedge STV_j[t_2] = 1$. This is a sufficient condition for the STVs not to cover one another, and, consequently, to guarantee independence.* ■

Lemma 3.5 *A test set MTV formed by independent STVs is aliasing-free for (a) S-A-0/1 and S-D-k faults, and (b) W-A/O faults.*

Proof: *Clause (a) follows from Lemma 3.2. To prove clause (b), only the W-A* *case will be proven; the W-O case can be proved by duality. For a W-A short S it is true that, due to STV independence,* $\forall n_i \in S : SRV_S \subset STV_i$. *Thus, for any W-A short S,* $SRV_S \notin MTV$, *which means W-A aliasing is impossible.* ■

Lemma 3.6 *In a test set MTV of independent and fixed weight STVs of weight W, any pair of disjoint faulty subsets* V_1/V_2 *cannot confound if they are characterized by a pair of different behaviors (except for S-A-0/W-A and S-A-1/W-O).*

Proof: *To prove this lemma one needs to prove two cases: (a) W-A/W-O, W-A/S-D-k and W-O/S-D-k, and (b) S-A-0/S-A-1, S-A-0/W-O, S-A-0/S-D-k, S-A-1/W-A, S-A-1/S-D-k, S-D-k/S-D-l. Case (a) is proved by noticing that* $H(SRV_{V_1}) \neq H(SRV_{V_2})$, *which implies* $SRV_{V_1} \neq SRV_{V_2}$. *Just note that, from STV independence and the fixed weight property,* $H(SRV_V) < W$ *if V is a W-A short,* $H(SRV_V) > W$ *if V is a W-O short and* $H(SRV_V) = W$ *if V is a S-D-k short. Case (b) follows from Lemma 3.3.* ■

Theorem 3.4 *The counting sequence plus complement test set MTV is (a) aliasing-free for S-A-0/1, W-A/O and S-D-k faults, and (b) confounding-free for pairs of different behaviors (except for S-A-0/W-A and S-A-1/W-O).*

Proof: *Clause (a) follows from Lemma 3.5, since MTV is a set of independent STVs. In turn, STV independence follows from Lemma 3.4, since MTV is by definition a fixed weight diagnostic scheme, where* $W = \lceil log_2 N \rceil$. *Clause (b) follows from Lemma 3.6.* ■

The above theorem shows that the counting sequence plus complement is aliasing-free for the synthetic fault models considered. Albeit not designed for that purpose, the counting sequence plus complement is confounding-free for many behavior pairs. Despite that, what is usually highlighted in the literature is its proness to confounding with respect to W-A/W-A or W-O/W-O pairs. One reason for that is the fact that the published analyses do not usually go beyond those two combinations of behaviors. As a matter of fact, the counting sequence plus complement is also prone to S-A-0/W-A, S-A-1/W-O, S-A-0/S-A-0 and S-A-1/S-A-1 confounding. For all its proness to confounding, the counting sequence plus complement is very popular in the industry. This may be because confounding does not occur very often, or simply because the cost of the confounding-free alternatives is too high. The confounding problem is illustrated by the following example:

Example 3.6 *Referring to Example 3.5, suppose there is a set of suspect nets* $U = \{n_1, n_2, n_7, n_8\}$ *for which* $SRV_U = \underline{1}$. SRV_U *could be produced by two disjoint faulty subsets* $V_1 = \{n_1, n_8\}$ *and* $V_2 = \{n_2, n_7\}$ *if* V_1 *and* V_2 *were characterized by W-O behavior. Therefore, according to Definition 3.8,* SRV_U *is a confounding response.*

3.3.3 Walking sequences

To get rid of W-O/W-O confounding, Hassan et al. proposed a test set given by $MTV = I$, where I is the Boolean identity matrix (Hassan et al., 1988). This test set was called the *walking-1s sequence*, since each STV contains exactly a single 1 that walks successively from position 1 to N. The test vector length is obviously given by

$$p = N. \tag{3.5}$$

Example 3.7 *The walking-1s test set for an 8-net circuit is as follows:*

$$MTV = \begin{bmatrix} 1 & 0 & 0 & 0 & 0 & 0 & 0 & 0 \\ 0 & 1 & 0 & 0 & 0 & 0 & 0 & 0 \\ 0 & 0 & 1 & 0 & 0 & 0 & 0 & 0 \\ 0 & 0 & 0 & 1 & 0 & 0 & 0 & 0 \\ 0 & 0 & 0 & 0 & 1 & 0 & 0 & 0 \\ 0 & 0 & 0 & 0 & 0 & 1 & 0 & 0 \\ 0 & 0 & 0 & 0 & 0 & 0 & 1 & 0 \\ 0 & 0 & 0 & 0 & 0 & 0 & 0 & 1 \end{bmatrix}$$

Instead of logarithmic complexity, the test vector length of the walking-1s sequence has a linear complexity $O(N)$. Unfortunately, this often leads to an unacceptable overhead in BS testing. In fact, since in BS the PTVs are applied serially, the time complexity for applying the walking-1s sequence would be $O(N^2)$, which may exceed the test time budget or the memory capacity of the ATE system (Section 1.6). Nevertheless, a positive aspect of this sequence is its increased diagnostic capability.

An important tool in the analysis of walking sequences is the concept of *diagonal independence* (Jarwala and Yau, 1989). This theoretical property is defined as follows:

Definition 3.12 (Diagonally independent test set) *A test set MTV ($N \times p$, $p \geq N$) is said to be diagonally independent iff by successive permutations of rows and/or columns, it is possible to obtain a test set MTV' such that*

$$STV_i'[t] = \begin{cases} 1, & i = t \\ 0, & i > t \\ X, & i < t \end{cases}$$

where 'X' stands for a don't care bit.

Note that changing the order of rows corresponds to changing the net labels in the circuit, which obviously does not alter the diagnostic capability. Changing the order of columns corresponds to changing the order of the PTVs, which does not alter the diagnostic capability either. Note that, being an interconnect circuit a memoryless circuit, it is not sensitive to the order in which the PTVs are applied. The general form of a diagonally independent test set is given by the following expression:

$$MTV = \begin{bmatrix} 1 & X & \dots & \dots & \dots & \dots & \dots & \dots \\ 0 & 1 & X & \dots & \dots & \dots & \dots & \dots \\ \dots & \dots & \dots & \dots & \dots & \dots & \dots & \dots \\ \dots & \dots & 0 & 1 & X & \dots & \dots & \dots \\ \dots & \dots & \dots & \dots & \dots & \dots & \dots & \dots \\ \dots & \dots & \dots & \dots & 0 & 1 & X & \dots \end{bmatrix}$$

According to Definition 3.12, it can be inferred that a diagonally independent test set is a set of unique STVs. Additionally, if STV_1 is chosen such that $STV_1 \neq \underline{1}$, the resulting test set is a diagnostic scheme according to Definition 3.5. The properties of diagonally independent diagnostic schemes are established by the following lemmas and theorem given below:

Lemma 3.7 *A diagonally independent diagnostic scheme MTV is aliasing-free for (a) S-A-0/1 and S-D-k faults, and for (b) W-A/O faults.*

Proof: *Clause (a) follows from Lemma 3.2. For clause (b) consider the W-O case; the W-A case is dual. Suppose there is a W-O short S that aliases a net* n_i*. This means that* $SRV_S = STV_i$*, and that it is possible to produce* SRV_S *without* n_i*. However, no nets* n_j*,* $j < i$*, can be used to produce* SRV_S *because they would make* $SRV_S[j] = 1 \neq STV_i[j] = 0$*. On the other hand, using solely nets* n_k*,* $k > i$*, will not do either, since they would make* $SRV_S[i] = 0 \neq STV_i[i] = 1$*. Thence, it is impossible to produce* SRV_S *without* n_i*, contrary to what has been asserted before. This leads to the conclusion that W-A/O faults are aliasing-free.* ■

Lemma 3.8 *In a diagonally independent diagnostic scheme, any pair of disjoint faulty subsets* V_1/V_2 *cannot confound if they are characterized by (a) a pair of different behaviors (except S-A-0/W-A, S-A-1/W-O and W-A/W-O) or (b) the pair of behaviors W-O/W-O.*

Proof: *To prove clause (a) one needs to prove cases (aa) S-A-0/S-A-1, S-A-0/W-O, S-A-0/S-D-k, S-A-1/W-A, S-A-1/S-D-k and S-D-k/S-D-l, and (ab) W-A/S-D-k and W-O/S-D-k. Case (aa) follows from Lemma 3.2. For case (ab) the W-O/S-D-k case is proven; the W-A/S-D-k case can be proved by duality. For W-O/S-D-k, no nets* n_i*,* $i < k$*, can be in* V_1 *or otherwise* $SRV_{V_1}[i] = 1 \neq SRV_{V_2}[i] = 0$*; also, a set* V_1 *consisting exclusively of the nets* n_i*,* $i > k$*, gives* $SRV_{V_1}[k] = 0 \neq SRV_{V_2}[k] = 1$*. Therefore, it is impossible to have W-O/S-D-k confounding. Case (ac) follows from Lemma 3.3.*

To prove clause (b) suppose V_1 *and* V_2 *can confound, so that* $SRV_{V_1} = SRV_{V2}$*. Consider the bit position i such that* $SRV_{V_1}[i] = 1 \wedge SRV_{V_1}[t] = 0$*,* $t < i$*. Then, it can immediately be concluded that* $n_i \in V_1$*, since it is impossible to reproduce* SRV_S *without* n_i*. But, because* $SRV_{V_1} = SRV_{V_2}$*, it is also concluded that* $n_i \in V_2$*. This goes against the initial assumption that* V_1 *and* V_2 *are disjoint, and the lemma has been proved by* reductio ad absurdum. ■

Theorem 3.5 *The walking-1s test set MTV is (a) aliasing-free for S-A-0/1, W-A/O and S-D-k faults, and (b) confounding-free for all behavior pairs but W-A/W-A, S-A-0/W-A, S-A-0/S-A-0 and S-A-1/S-A-1.*

Proof: *Clause (a) follows from Lemma 3.7, since MTV is diagonally independent according to Definition 3.12. For clause (b) the cases (ba) S-A-1/W-O and (bb) W-A/W-O must be proved; the other cases follow from Lemma 3.8. For case (ba), to produce* $SRV_{V_2} = \underline{1}$ *one needs all STVs in the test set. Because* V_1 *is non-empty,* V_2 *cannot contain all the N nets, which means* $STV_{V_2} = \underline{1}$ *cannot be produced. Thus,* $SRV_{V_2} \neq SRV_{V_1}$*, and S-A-1/W-O pairs do not confound. In case (bb),* $SRV_{V_1} = \underline{0} \wedge SRV_{V_2} \neq \underline{1}$*. Thus* $SRV_{V_1} \neq SRV_{V_2}$*, and W-A/W-O pairs do not confound.* ∎

An alternative way to prove aliasing-freedom for the synthetic fault models and confounding-freedom for all but S-A-0/W-A, S-A-1/W-O W-A/W-A and W-O/W-O behavior pairs is by using the fact that MTV is a set of unique STVs of fixed weight $W = 1$. This implies STV independence, which, in turn, produces the proof using Lemmas 3.5 and 3.6. This is the process used in Theorem 3.4 for the counting sequence plus complement. Instead, Theorem 3.5 uses Lemmas 3.7 and 3.8, which will be necessary to prove the properties of two other diagonally independent schemes.

The diagonal independence concept introduced by Definition 3.12 has permitted the W-O/W-O confounding problem to be solved. Swapping 0s and 1s in Definition 3.12 results in a dual concept of diagonal independence which can solve the W-A/W-A confounding problem. The diagonal independence property as introduced by Definition 3.12 will be called W-O diagonal independence, and its dual will be called W-A diagonal independence. The walking-1s sequence is a W-O diagonally independent test set and its dual, the *walking-0s sequence*, is a W-A diagonally independent test set.

To address both W-O/W-O and W-A/W-A confounding, Hassan et al. also proposed the test set $MTV = [I\ \overline{I}]$ (Hassan et al., 1988). This diagnostic scheme can be thought of as a walking-1s followed by a walking-0s sequence, and can be called the *complete walking sequence*. The STV length is obviously given by

$$p = 2N. \tag{3.6}$$

The diagnostic capability of the complete walking sequence subsumes the capabilities of the walking-1s and walking-0s sequences. It is formally established by the following theorem:

Theorem 3.6 *The complete walking sequence test set MTV is (a) aliasing-free for S-A-0/1, W-A/O and S-D-k faults, and (b) confounding-free for all but the S-A-0/S-A-0 and the S-A-1/S-A-1 behavior pairs.*

Proof: *Clause (a) follows from Lemma 3.7, since MTV is W-O diagonally independent. Note that MTV matches Definition 3.12. Clause (b) follows from* *the fact that MTV is both W-O and W-A diagonally independent.* [4] *Therefore, this clause can be proved using a process similar to that used for Clause (b) of Theorem 3.5.* ■

The confounding pairs S-A-0/S-A-0 and S-A-1/S-A-1 cannot be solved at all by any behavioral test set. Just note that the response of S-A-0/1 faults does not depend on the involved STVs; it is imposed by ground and power nets. Since the problem cannot be solved by any further optimizations of the test set, and since it is the only ambiguity left, it can be concluded that the complete walking sequence achieves minimal diagnostic ambiguity.

3.3.4 Min-weight sequence

The min-weight sequence (Yau and Jarwala, 1989) has the original feature of providing tunable diagnostic capability. It is obtained by minimizing the Hamming weight of the STVs for a given STV length p, while ensuring that there are enough STVs for all nets. The value of p can be selected in the range $\lceil \log(N+2) \rceil \leq p \leq N$. For $p = \lceil \log(N+2) \rceil$, the lower limit of the range, the test set is equivalent to the modified counting sequence (Section 3.3.1); for $p = N$, the upper limit in the range, the test set is equivalent to the walking-1s test set (Section 3.3.3). By choosing an appropriate value of p, a good compromise between STV length and diagnostic capability can be achieved.

Example 3.8 *For an 8-net circuit and* $p = 4$, *the min-weight is given as follows:*

$$MTV = \begin{bmatrix} 0 & 0 & 0 & 1 \\ 0 & 0 & 1 & 0 \\ 0 & 1 & 0 & 0 \\ 1 & 0 & 0 & 0 \\ 0 & 0 & 1 & 1 \\ 0 & 1 & 0 & 1 \\ 1 & 0 & 0 & 1 \\ 0 & 1 & 1 & 0 \end{bmatrix}$$

The working principle of the min-weight sequence is simple: minimize the STV Hamming weight to minimize the probability of W-O related aliasing and confounding. Since the number of unique STVs of Hamming weight w is $\binom{p}{w}$, it is interesting to note that the maximum Hamming weight W_{max} present in a W-O min-weight test set, is given by

$$W_{max} = \min W, \qquad \text{such that} \qquad \sum_{w=1}^{W} \binom{p}{w} \geq N. \tag{3.7}$$

[4] To see that the complete walking sequence is also W-A diagonally independent, note that I and $\overline{I}$ can be permuted in MTV, which produces a test set $MTV' = [\overline{I}\ I]$ that matches the dual of Definition 3.12.

An aliasing-free test set with symmetric diagnostic capability for W-O and W-A shorts can be obtained by concatenating the min-weight sequence with its complement (Rogel-Favila, 1992). The test set thus obtained may be called the ***min-weight sequence plus complement***, and has an STV length in the range $2\lceil\log(N+2)\rceil \leq p \leq 2N$. For $p = 2\lceil\log(N+2)\rceil$, the lower limit in this range, the min-weight sequence plus complement is equivalent to the modified counting sequence plus complement (Section 3.3.2). For $p = 2N$, the upper limit in the range, the min-weight sequence plus complement corresponds to the complete walking sequence (Section 3.3.3).

3.3.5 Self-diagnosis sequence

An important property of test sets of independent STVs is the ***self-diagnosability*** property with respect to W-A/O and S-A-0/1 behaviors, which is defined as follows (Cheng et al., 1990):

Definition 3.13 (Self-diagnosable test set) *A test set is said to be self-diagnosable for a particular fault behavior iff it makes any faulty net affected by that behavior produce an erroneous response.*

Self-diagnosability for a certain behavior implies aliasing-freedom for that behavior, and is actually a sufficient condition for aliasing-freedom — a net that produces an erroneous response is always faulty. However, it is not a necessary condition. For example, in a S-D-k short the strong-driver net n_k is not self-diagnosable because it always produces a correct response. Nevertheless, according to Lemma 3.2, S-D-k shorts never cause aliasing in any behavioral diagnostic scheme. Examples of aliasing-free W-O shorts containing non self-diagnosable nets could also be given. To produce an optimal aliasing-free test set, a sufficient and necessary condition for aliasing-freedom must be identified. Since self-diagnosability is not a necessary condition for aliasing-freedom, this concept cannot produce optimal aliasing-free test sets.

From the proof of Lemma 3.5, it follows that STV independence implies S-A-0/1 and W-A/O self-diagnosability: in a diagnostic scheme of independent STVs, faulty nets having S-A-0/1 or W-A/O behaviors always produce erroneous responses. Moreover, Lemma 3.5 establishes that STV independence guarantees aliasing-freedom for the synthetic fault models considered. Since the counting sequence plus complement is a set of independent STVs, it is aliasing-free for all synthetic behaviors (Section 3.3.2). However, the counting sequence plus complement is not the optimal set of independent STVs in terms of test vector length. The optimal set of independent STVs is obtained by generating all vectors of length p and Hamming weight $W = \lfloor p/2 \rfloor$ or $W = \lceil p/2 \rceil$. Such test set has been proposed as an improvement on the counting sequence plus complement, and has been called the ***self-diagnosis sequence*** because it is self-diagnosable for W-A/O shorts (Cheng et al., 1990). Its test vector length is given by

$$p = \min k, \qquad \text{such that} \qquad \binom{k}{\lfloor \frac{k}{2} \rfloor} \geq N. \tag{3.8}$$

It is possible to show that the value of p given by the above equation is in the range

$$\lceil \log_2 N \rceil < p < 2\lceil \log_2 N \rceil, \tag{3.9}$$

which means that the STV length of the self-diagnosis sequence is lower than the STV length of the counting sequence plus complement (Equation (3.4)) but higher than the STV length of the counting sequence (Equation (3.1)).

Example 3.9 *The self-diagnosis sequence for an 8-net circuit is as follows:*

$$MTV = \begin{bmatrix} 0 & 0 & 0 & 1 & 1 \\ 0 & 0 & 1 & 0 & 1 \\ 0 & 0 & 1 & 1 & 0 \\ 0 & 1 & 0 & 0 & 1 \\ 0 & 1 & 0 & 1 & 0 \\ 0 & 1 & 1 & 0 & 0 \\ 1 & 0 & 0 & 0 & 1 \\ 1 & 0 & 0 & 1 & 0 \end{bmatrix}$$

The self-diagnosis sequence is a set of unique, independent and fixed weight STVs ($W = \lfloor p/2 \rfloor$). Therefore, its diagnostic capability is the same as that of the counting sequence plus complement (Theorem 3.4).

As discussed in the literature (Cheng et al., 1990), there is a similarity between behavioral diagnosis test sets and the theory of *asymmetric error detecting codes* (Bose and Rao, 1982). Shorts correspond to *asymmetric errors*, and the self-diagnosis capability corresponds to *all error detecting capability.* The counting sequence plus complement corresponds to *$p/2$-pair two-rail codes*, and the self-diagnosis test set corresponds to *$p/2$-out-of-p codes*, which can be proved to be optimal (Freiman, 1962).

3.3.6 Global-diagnosis sequence

The test set of the previous section could be further improved by identifying the property that provides a necessary and sufficient condition for aliasing-freedom. This property is known as *global diagnosability* (Cheng et al., 1990). A diagnostic scheme that possesses this property is called a *globally diagnosable* test set and is defined as follows:

Definition 3.14 (Globally diagnosable test set) *A test set is said globally diagnosable with respect to a certain fault behavior iff it makes any faulty net affected by that behavior produce a response which is identical to the response of some net that produces an erroneous response.*

A globally diagnosable test set does not require that a faulty net n_i produces an erroneous response SRV_i. As long as there is some net n_j that produces an erroneous response $SRV_j = SRV_i$, it can be concluded that n_i is faulty. In this way, it is always possible to identify faulty nets, which means that global-diagnosability implies aliasing-freedom.

Nets can be self-diagnosable with respect to some fault behaviors and globally diagnosable with respect to others. In the previous section is has been shown that the aliasing-free methods presented so far are self-diagnosable for S-A-0/1 and W-A/O faults. It has also been shown that these methods are aliasing-free, albeit not self-diagnosable for S-D-k shorts. In fact, any behavioral diagnostic scheme is globally diagnosable with respect to S-D-k shorts: the strong-driver net n_k can be diagnosed while producing a correct response.

Cheng et al. proposed a test set which provides global-diagnosability with respect to W-O shorts (Cheng et al., 1990). This test set was called the *global-diagnosis sequence*, and is given recursively by the following equations:

$$\begin{cases} MTV^{SD}(1) = [0] \\ MTV^{GD}(1) = [0\ 1]^T \\ MTV^{GD}(p+1) = \begin{bmatrix} [MTV^{GD}(p)\ \underline{0}^T] \\ [MTV^{SD}(p)\ \underline{1}^T] \end{bmatrix} \end{cases} \tag{3.10}$$

where $MTV^{SD}(p)$ and $MTV^{GD}(p)$ are the self-diagnosis and the global-diagnosis sequences, respectively, for an STV length p. For a global-diagnosis sequence, p is given by

$$p = \min k, \qquad \text{such that} \qquad 2 + \sum_{i=1}^{k-1} \binom{i}{\lfloor \frac{i}{2} \rfloor} \geq N,\ N > 2. \tag{3.11}$$

It is possible to show that p, as given by the above expression, satisfies the inequalities

$$\lceil \log_2 N \rceil < p < p^{SD}, \tag{3.12}$$

where p^{SD} is the STV length of the self-diagnosis test set (Equation (3.8)). That is, the STV length of the global-diagnosis sequence is higher than the STV length of the counting sequence (Equation (3.1)) but lower than that of the self-diagnosis sequence (Equation (3.8)). This improvement over the self-diagnosis sequence is possible because the global-diagnosability criterion requires fewer vectors than the self-diagnosability one.

Example 3.10 *The global-diagnosis sequence for an 8-net circuit is as follows:*

$$MTV = \left[\begin{array}{ccc|c} 0 & 0 & 0 & 0 \\ 1 & 0 & 0 & 0 \\ 0 & 1 & 0 & 0 \\ 0 & 1 & 1 & 0 \\ 1 & 0 & 1 & 0 \\ \hline 0 & 0 & 1 & 1 \\ 0 & 1 & 0 & 1 \\ 1 & 0 & 0 & 1 \end{array}\right]$$

The dividing lines are tell tale of the recursive technique expressed by Equation (3.10), which is used to build the sequence.

Before proving the diagnostic properties of the global-diagnosis sequence, it is necessary to show that the test set is a diagnostic scheme. According to Definition 3.5, one needs to show that the STVs in the sequence are unique and do not contain the vectors $\underline{0}$ and $\underline{1}$. Since it is not obvious from Equation (3.10) that the global-diagnosis sequence is a set of unique STVs, the following lemma provides the proof:

Lemma 3.9 *The global-diagnosis sequence is a set of unique STVs.*

Proof: *Induction is used. From Equation (3.10) $MTV^{GD}(1)$ is a set of unique STVs. Now one has to prove that if $MTV^{GD}(p)$ is a set of unique STVs then $MTV^{GD}(p+1)$ is also a set of unique STVs. Because $MTV^{GD}(p)$ and $MTV^{SD}(p)$ are sets of unique STVs, $[MTV^{GD}(p)\ \underline{0}^T]$ and $[MTV^{SD}(p)\ \underline{1}^T]$ are also sets of unique STVs. Moreover, the STVs in $[MTV^{GD}(p)\ \underline{0}^T]$ are different from those in $[MTV^{SD}(p)\ \underline{1}^T]$ because they differ in the last bit. Therefore, every STV in $MTV^{GD}(p+1)$ is unique, which concludes the proof by induction.*

■

The global-diagnosis sequence as given by Equation (3.10) is not a diagnostic scheme because it contains the vector $STV_1 = \underline{0}$. To make it a diagnostic scheme, this vector must be excluded from MTV, which slightly increases p to

$$p = \min k, \qquad \text{such that} \qquad 1 + \sum_{i=1}^{k-1} \binom{i}{\lfloor \frac{i}{2} \rfloor} \geq N,\ N > 2. \tag{3.13}$$

Having shown how to obtain a diagnostic scheme from the global-diagnosis sequence, one can now proceed to establish its diagnostic capability by means of Theorem 3.7 below. Clause (b) in Theorem 3.7, which concerns W-O aliasing-freedom, is basically reproduced from the literature (Cheng et al., 1990) with some added details.

Theorem 3.7 *The global-diagnosis scheme is (a) aliasing-free for S-A-0/1 and S-D-k faults, (b) aliasing-free for W-O shorts and (c) confounding-free for S-A-0/S-A-1, S-A-0/W-O, S-A-0/S-D-k, S-A-1/W-A, S-A-1/S-D-k and S-D-k/S-D-l behavior pairs.*

Proof: *Clause (a) follows from Lemma 3.2. Clause (b) is proved using induction. From Equation (3.10) it is clear that $MTV^{GD}(1)$ is aliasing-free for W-O shorts. Now one needs to prove that if $MTV^{GD}(p)$ is aliasing-free for W-O shorts, so is $MTV^{GD}(p+1)$. Since $MTV^{GD}(p)$ and $MTV^{SD}(p)$ are aliasing-free for W-O shorts, so are $[MTV^{GD}(p)\ \underline{0}^T]$ and $[MTV^{SD}(p)\ \underline{1}^T]$. The proof is completed by showing that (ba) a W-O short S involving the nets in $[MTV^{GD}(p)\ \underline{0}^T]$ does not produce aliased nets n_j in $[MTV^{SD}(p)\ \underline{1}^T]$, (bb) a short S involving at least a net n_j in $[MTV^{SD}(p)\ \underline{1}^T]$ does not produce aliased nets n_i in $[MTV^{GD}(p)\ \underline{0}^T]$, and (bc) a short S involving at least a net n_j in $[MTV^{SD}(p)\ \underline{1}^T]$ does not produce aliased nets in $[MTV^{SD}(p)\ \underline{1}^T]$. Case (ba) follows from the fact that one would have $SRV_S[p+1] = 0 \neq STV_j[p+1] = 1$. Case (bb) follows from the fact that one would have $SRV_S[p+1] = 1 \neq$*

$STV_i[p+1] = 0$. Case (bc) follows from two facts. First, if $n_j \in S$ is the only net in $[MTV^{SD}(p)\ \underline{1}^T]$, it cannot be aliased because without it one could not have $SRV_S[p+1] = 1$. Second, if n_j is not the only net in $[MTV^{SD}(p)\ \underline{1}^T]$, then $SRV_S \supset STV_j$ because the STVs in $[MTV^{SD}(p)\ \underline{1}^T]$ are independent by definition. Thence, $\forall STV_j \in [MTV^{SD}(p)\ \underline{1}^T] : STV_j \neq SRV_S$, and no net can be aliased in $[MTV^{SD}(p)\ \underline{1}^T]$. Finally, clause (c) follows from Lemma 3.3. ∎

The global-diagnosis sequence achieves aliasing-freedom for W-O shorts while having shorter STVs than the self-diagnosis sequence. However, unlike the self-diagnosis scheme, which has symmetric capability for W-A and W-O shorts, the global-diagnosis sequence has been designed for W-O shorts only. It can easily be shown that its diagnostic capability for W-A shorts is rather weak.

Using duality, it is possible to construct a W-A aliasing-free global diagnosis sequence. However, it is unknown how to construct a W-A/O symmetric global-diagnosis sequence that has aliasing-free capability and an STV length shorter than the self-diagnosis sequence. It is possible to construct an aliasing-free scheme by concatenating the global-diagnosis sequence and its complement. That produces a test of independent STVs like the counting sequence plus complement or the self-diagnosis sequence. However, such test set would have longer STVs than both those schemes, as can be concluded from Equation (3.11), Inequality (3.9) and Inequality (3.12).

Due to its limited diagnostic capability, the global-diagnosis test set is of little practical use. Despite that, global-diagnosability is a powerful concept, which is fully exploited in the test set of the next section.

3.3.7 N+1 sequence

The complete walking sequence of Section 3.3.3 was for a long time the only known scheme capable of fully minimizing diagnostic ambiguity. Unfortunately, its STV length $p = 2N$ (Equation (3.6)) may be excessive for BS application. A scheme also having minimal diagnostic ambiguity but having an STV length given by

$$p = N + 1 \tag{3.14}$$

has been proposed by Park, who called this scheme the *N+1 sequence* (Park, 1996). The test set is given by $MTV = [\underline{0}^T\ UT]$, where UT is an $N \times N$ *upper triangular* Boolean matrix. [5]

[5] An upper triangular Boolean matrix has elements which are 1 if they are in or above the main diagonal, and 0 otherwise.

Example 3.11 *For an 8-net circuit, the N+1 sequence is given by the following expression:*

$$MTV = \begin{bmatrix} 0 & 1 & 1 & 1 & 1 & 1 & 1 & 1 & 1 \\ 0 & 0 & 1 & 1 & 1 & 1 & 1 & 1 & 1 \\ 0 & 0 & 0 & 1 & 1 & 1 & 1 & 1 & 1 \\ 0 & 0 & 0 & 0 & 1 & 1 & 1 & 1 & 1 \\ 0 & 0 & 0 & 0 & 0 & 1 & 1 & 1 & 1 \\ 0 & 0 & 0 & 0 & 0 & 0 & 1 & 1 & 1 \\ 0 & 0 & 0 & 0 & 0 & 0 & 0 & 1 & 1 \\ 0 & 0 & 0 & 0 & 0 & 0 & 0 & 0 & 1 \end{bmatrix}$$

The diagnostic capability of the N+1 sequence is established by the following theorem:

Theorem 3.8 *The N+1 sequence is (a) aliasing-free for S-A-0/1, W-A/O and S-D-k faults, and (b) confounding-free for all but the S-A-0/S-A-0 and the S-A-1/S-A-1 behavior pairs.*

Proof: *Clause (a) follows from Lemma 3.7, since MTV is W-O diagonally independent. To show that MTV is W-O diagonally independent, it is shifted circularly one bit to the left to obtain a matrix MTV' that matches Definition 3.12.*

To prove clause (b), it is necessary to show that MTV is also W-A diagonally independent. This is accomplished by flipping it horizontally, then vertically, and finally shifting it circularly one bit to the left. [6] *A matrix MTV' that matches the dual of Definition 3.12 is obtained in this way. Now, let clause (b) be divided in the cases (ba) S-A-1/W-O and S-A-0/W-A, and (bb) W-A/W-O; all other cases follow from Lemma 3.8 and its dual. For case (ba), the S-A-1/W-O part is proven; the S-A-0/W-A part can be proved by duality. $SRV_{V_2} = \underline{1}$ cannot be produced since $\forall i : STV_i[1] = 0$. Thus $SRV_{V_2} \neq SRV_{V_1}$, and S-A-1/W-O pairs do not confound. For case (bb), suppose there is W-A/W-O confounding so that $SRV_{V_1} = SRV_{V_2}$. Let n_i be such that $SRV_S[t] = 0,\ t < i$, and $SRV_S[t] = 1,\ t \geq i$. If V_1 is a W-A faulty subset, then $n_i \in V_1$ because it is impossible to produce $SRV_{V_1} = SRV_S$ without n_i. Similarly, if V_2 is a W-O faulty subset, it is impossible to produce $SRV_{V_2} = SRV_S$ without n_i. But, n_i cannot simultaneously be in V_1 and V_2 because that goes against the hypothesis that V_1 and V_2 are disjoint. Thence, clause (bb) has been proven by* reductio ad absurdum. ■

In addition to being globally diagnosable with respect to S-D-k shorts, in the N+1 sequence, nets are also globally diagnosable with respect to W-A and W-O shorts. Note that, in a W-A (W-O) short S, the net n_i with the highest

[6] Shifting and flipping operations are equivalent to row/column permutations, as can easily be shown.

(lowest) subscript i always produces a correct response $SRV_i = SRV_S = STV_i$. Nonetheless, n_i must be faulty, since, according to Clause (b) in Lemma 3.7, it is impossible to explain SRV_S otherwise. Hence, in any W-A/O or S-D-k short S there is a globally diagnosable net n_i. This fact automatically guarantees confounding-freedom, since two distinct shorts have two distinct globally-diagnosable nets n_1 and n_2. These nets are associated with two disjoint faulty subsets V_1 and V_2, which produce two distinct responses $SRV_{V_1} = STV_1 \neq SRV_{V_2} = STV_2$. An alternative proof of the theorem above that uses the concept of global-diagnosability has just been outlined. The global-diagnosability property with respect to all but the S-A-0/1 behavior is the characteristic that yields the compactness and minimal diagnostic ambiguity of the N+1 sequence.

3.3.8 N sequence

In proving Theorem 3.8 recourse has been made to the fact that the N+1 sequence is both W-A and W-O diagonally independent. However, this test set is not *the optimal* W-A and W-O diagonally independent test set. The optimal W-A and W-O diagonally independent test set (de Sousa, 1998) is given by an $N \times N$ matrix defined by

$$STV_i'[t] = \begin{cases} 1, & i = t \ \vee (i < t \ \wedge \ t \neq N) \\ 0, & i > t \ \vee \ (i = 1 \ \wedge \ t = N) \\ X, & \text{otherwise} \end{cases}$$

The general form of this matrix is

$$MTV = \begin{bmatrix} 1 & 1 & \dots & \dots & \dots & \dots & 0 \\ 0 & 1 & 1 & \dots & \dots & \dots & X \\ \dots & \dots & \dots & \dots & \dots & \dots & \dots \\ \dots & \dots & 0 & 1 & 1 & \dots & \dots \\ \dots & \dots & \dots & \dots & \dots & \dots & \dots \\ \dots & \dots & \dots & \dots & 0 & 1 & X \\ \dots & \dots & \dots & \dots & \dots & 0 & 1 \end{bmatrix}$$

and the STV length p is obviously given by

$$p = N. \tag{3.15}$$

The diagnostic capability of the N sequence is established by the following lemma and theorem:

Lemma 3.10 *The N sequence test set is W-A and W-O diagonally independent, and is optimal.*

Proof: *MTV is W-O diagonally independent because it matches Definition 3.12. To see it is also W-A diagonally independent, shift the matrix circularly one bit to the right, and then flip it both horizontal and vertically. In this way, a matrix MTV' that matches the dual of Definition 3.12 is obtained. This test* *set is optimal because, since $p \geq N$, then $p = N$ is the shortest possible STV length for an $N \times p$ diagonally independent matrix.* ■

Theorem 3.9 *The N sequence test set is (a) aliasing-free for S-A-0/1, W-A/O and S-D-k faults, and (b) confounding-free for all but the S-A-0/W-A, S-A-1/W-O, S-A-0/S-A-0 and the S-A-1/S-A-1 behavior pairs.*

Proof: *Clause (a) follows from Lemma 3.7, since MTV is W-A/O diagonally independent according to Lemma 3.10. In clause (b) all cases follow from Lemma 3.8 and its dual, except the W-A/W-O case which can be proven like clause (bb) in Theorem 3.8.* ■

The fact that the N sequence cannot resolve S-A-0/W-A and S-A-1/W-O confounding pairs (examples could easily be given) evidences the weakness of the traditional analyses based solely on W-A/W-A and W-O/W-O confounding. Such analyses are unable to distinguish between the N and the N+1 sequences, whereas the analysis presented here can. In fact, since S-A-0/1 faults are quite common, and since the STV lengths $p = N$ and $p = N + 1$ are practically equivalent, in a real situation one would always prefer the N+1 sequence rather than the N sequence.

3.4 Summary

In this chapter, an original analysis of behavioral interconnect diagnosis has been presented. It is assumed that any net can have an open fault and that any two nets can be involved in a short fault. These general assumptions enable diagnosis to be performed without information on the layout or potential faults. The following issues have been discussed: diagnostic analysis, fault isolation, and diagnostic synthesis.

A diagnostic analysis procedure that draws from the analytic fault models studied in Chapter 2 has been presented. This procedure can be summarized as follows: nets producing identical responses are potentially shorted; nets producing erroneous responses or the same response as a net that produces an erroneous response are potentially open; nets producing $\underline{0}/\underline{1}$ responses are potentially shorted to $n_{G/P}$ nets. This approach is meant to address the difficulties in predicting the behavior of realistic faults. It also simplifies diagnostic analysis to $O(N \log N)$ time complexity. Diagnoses have been classified as *complete, incomplete* or *incorrect*, and, additionally, as *ambiguous, unambiguous* or *perfect*. This classification is used to evaluate the effectiveness of diagnostic schemes.

Fault isolation is the procedure of physically locating faults in the EA using the diagnostic information produced during response analysis. This procedure must be carefully conducted to be able to deal with the problem of *fault masking*. The way it accomplishes that is by first isolating the directly observable faults, then repairing them, and after retesting the EA. Although a diagnosis may contain potentially masked faults, they should be ignored until the masking faults are repaired. Then, the previously hidden faults become observable

Table 3.1. Diagnostic properties of behavioral schemes.

Scheme	Length p	Aliasing-free?					Confounding-free?					
		S-A-0	S-A-1	W-A	W-O	S-D-k	V_1, V_2	S-A-0	S-A-1	W-A	W-O	S-D-k
Modified Counting Sequence	$\lceil \log_2(N+2) \rceil$	yes	yes	no	no	yes	S-A-0	no	yes	no	yes	yes
							S-A-1	—	no	yes	no	yes
							W-A	—	—	no	no	no
							W-O	—	—	—	no	no
							S-A-l	—	—	—	—	yes
Self Diagnosis Sequence	$\lceil \log_2 N \rceil < p < 2\lceil \log_2 N \rceil$	yes	yes	**YES**	**YES**	yes	S-A-0	no	yes	no	yes	yes
							S-A-1	—	no	yes	no	yes
							W-A	—	—	no	**YES**	**YES**
							W-O	—	—	—	no	**YES**
							S-A-l	—	—	—	—	yes
N Sequence	N	yes	yes	yes	yes	yes	S-A-0	no	yes	no	yes	yes
							S-A-1	—	no	yes	no	yes
							W-A	—	—	**YES**	yes	yes
							W-O	—	—	—	**YES**	yes
							S-A-l	—	—	—	—	yes
N+1 Sequence	$N+1$	yes	yes	yes	yes	yes	S-A-0	no	yes	**YES**	yes	yes
							S-A-1	—	no	yes	**YES**	yes
							W-A	—	—	yes	yes	yes
							W-O	—	—	—	yes	yes
							S-A-l	—	—	—	—	yes

and can be isolated and repaired. This procedure avoids wasting time looking for faults before one knows they exist.

The diagnostic capability of the most important behavioral interconnect diagnostic schemes has been studied in detail. Mixed short-open faults of any multiplicity whose components behave according to the S-A-0/1, W-A/O and S-D-k synthetic fault models are considered. The diagnostic capability of the various schemes is analyzed based on the well known aliasing/confounding framework. These properties are evaluated for any combination of behaviors of the component faults in a multiple mixed fault.

A new test set has been derived. It has been called the N sequence, inspired by Park's N+1 sequence. As the name indicates, it produces STVs of length N, which is only one bit less than the N+1 sequence. However, the interesting aspect of this sequence is its optimality as a W-A/O diagonally independent test set. From the analysis of the diagnostic properties of the N sequence it has been concluded that, for the fault models considered, the W-A/O diagonally independence property is not a sufficient condition for achieving minimal diagnostic ambiguity.

The diagnostic capability of four representative behavioral schemes is summarized in Table 3.1. For each scheme, the test vector length p and "yes/no" indications of aliasing-freedom and confounding-freedom are given. The indication of aliasing-freedom is given for all synthetic fault models considered, and the indication of confounding-freedom is given for all possible pairs of behaviors. The representative schemes are: the modified counting sequence, the self-diagnosis sequence, the N sequence and the N+1 sequence. The schemes are ordered by increasing diagnostic capability. When "**YES**" appears instead of "yes", it means that the respective diagnostic capability cannot be offered by the preceding schemes. In this way, the following facts are highlighted: the self-diagnosis sequence can avoid aliasing responses for W-A/O behaviors and confounding responses for W-A/W-O, W-A/S-D-k and W-O/S-D-k behavior pairs; the N sequence can prevent W-A/W-A and W-O/W-O confounding responses; the N+1 sequence can resolve S-A-0/W-A and S-A-1/W-O confounding pairs.

Traditional analysis restricted to W-A/O faults cannot distinguish between the N and the N+1 sequence because they do not consider S-A-0/W-A and S-A-1/W-O confounding pairs. However, since S-A-0/1 faults may be quite common and the two sequences have approximately the same length, the N+1 sequence is a better choice than the N sequence. S-D-k shorts do not pose any additional problems to behavioral schemes, which could have been derived without taking this behavior into account. In the next chapter it will be shown that in structural diagnosis S-D-k shorts must be explicitly considered, or otherwise the diagnostic capability is significantly affected.

4 STRUCTURAL INTERCONNECT DIAGNOSIS

In this chapter an original study of *structural* diagnosis of interconnect faults is presented. Structural interconnect diagnosis results from combining the diagnostic sequences studied in the previous chapter with the knowledge of possible layout fault locations. This knowledge enables further optimization of diagnostic schemes, both in serial test vector (STV) length and diagnostic capability. The new perspective on structural diagnosis introduced in this work has led to the development of innovative methods that exploit defect statistics.

Because structural diagnosis depends on the layout, more complex algorithms are needed when compared with behavioral diagnosis. Nevertheless, due to the increasing complexity of electronic assemblies (EAs), structural methods seem the most plausible way to restrict the complexity of diagnostic schemes while improving diagnostic capability. Additionally, this work aims at establishing foundations for automatic repair technology. With this goal in mind, the inclusion of structural information is inevitable.

Few structural diagnosis schemes for interconnects have been proposed in the literature. Main limitations of those schemes reside in the fault models used, and the incomplete treatment of diagnostic capability. The schemes proposed in this research tackle both these limitations. They benefit from the more comprehensive fault modeling approach presented in Chapter 2, and from a rigorous treatment of diagnostic capability, backed by statistical information on defects.

The chapter follows the organization of the previous chapter, except that an extra section, Section 4.1, is included to deal with the issue of layout fault extraction. Diagnostic analysis is discussed in Section 4.2, fault isolation in Section 4.3, and diagnostic synthesis in Section 4.4. Section 4.5 presents a summary of the chapter.

4.1 Extraction of faults

In order to have an experimental setup for fault extraction, a PCB fault extraction tool has been implemented. The tool is used to identify layout single shorts and layout single opens. Multiple faults are not directly extracted from the layout because that would lead to a computationally expensive problem. The diagnostic analysis and diagnostic synthesis programs will input single faults, and will implicitly enumerate multiple faults as being combinations of single faults.

The fault extraction procedure creates a mapping between each single fault and the physical locations of the defects that can cause the fault. This is a *one to many* relationship as each single fault may be associated with many possible physical defects. In structural diagnosis, all the defect locations are stored to be used later in fault isolation. The structural information is kept in a suitable data structure, which can also be used for automatic repair. Note that in behavioral diagnosis no structural information is available and, therefore, no automatic repair procedures can be built upon it.

4.1.1 Single opens

In PCBs, the main cause of open faults is missing solder accidentally disconnecting net receivers from their drivers (Wassink, 1989). It can be assumed that opens can only occur at solder joints, which are mainly located at component pins. Each net n_i is associated with a single open fault o_i, which, in turn, is associated with a set of physical defects.

In order to extract all possible opens for a net n_i, it must be possible to retrieve the locations of all its pins and other solder joints. In terms of data structure, the minimum requirement is that each layout object has an identifier of the net to which it belongs. Ideally, this data structure should be created when the layout itself is created to avoid the overhead of generating it from an already designed layout. Unfortunately, this is seldom the case. Normally, the data structure is created after the layout has been designed. This can be done by checking the connectivity among layout geometries, which is accomplished in $O(N \log N)$ time. The present discussion assumes that each layout object has already been tagged with a net identifier and, therefore, the work required to compute these data is not included in the complexity of the fault extraction procedure.

4.1.2 *Single shorts*

The problem of identifying the potential net shorts in a PCB layout is not novel (Cheng et al., 1990; Garey et al., 1976; Hedelsbrunner, 1987; Lewandowski and Velasco, 1986). In a PCB layout, nets may accidentally be involved in short circuits due to their physical proximity. Nets that can short directly are called *neighbors*:

Definition 4.1 (Neighbor nets) *Two nets n_i and n_j are said to be* neighbors *if a direct short $e_{i,j}$ between n_i and n_j is physically possible.*

Deciding whether two nets are neighbors is not a trivial problem, since for each technology different criteria may apply. Nevertheless, two technology independent criteria have been used:

- *Distance*: a single short $e_{i,j}$ is extracted if $\overline{PQ} < R$, where P and Q are some points from n_i and n_j, respectively, and R is the *maximum short radius.* R depends of course on the technology being used.
- *Visibility*: a short $e_{i,j}$ is extracted if it is possible to draw a line segment PQ from some point $P \in n_i$ to some point $Q \in n_j$ without intersecting any other net.

These criteria have been introduced to formulate the short extraction problem in bare PCBs. The distance and the visibility criteria can also be combined, so that a short $e_{i,j}$ is extracted only if n_i and n_j are visible from each other and $\overline{PQ} < R$. In assembled PCBs, the problem is much simpler since soldering is the last relevant fabrication step. The main cause of shorts is excess solder accidentally connecting two or more nets (Wassink, 1989). Because solder is used to bond component pins to their layout pads, it can be assumed that shorts will only occur in the vicinity of component pins.

Two extraction modes have been considered: the *pin-to-net* mode and the *pin-to-pin* mode. The pin-to-net mode is used in technologies where shorts may occur between a pin and any reachable neighbor geometry. The pin-to-pin mode is used in technologies where shorts can only occur between neighbor pins. [1]

The distance criterion is an efficient and easy to implement procedure. On the other hand, the visibility criterion can become computationally expensive, and difficult to implement. Hence, only the distance criterion has been used in this work. Fortunately, the undesirable effects of ignoring the visibility criterion are minor, as explained in the next two paragraphs.

In the pin-to-pin mode, the distance rule is enforced by restricting the short radius R bounded as $PS < R < 2PS$, where PS is the spacing between IC pins. In this case the visibility criterion is implicitly observed: pins that are invisible from each other are necessarily separated by a distance equal to or greater

[1] This is the case where the bare board is covered with a layer of solder resist, so that only the pin pads are exposed to solder.

than $2PS$. Hence, in the pin-to-pin mode, ignoring the visibility criterion has no consequences provided all the ICs have the same pin spacing PS. [2]

In the pin-to-net mode, layout geometries other than pins with different spacing rules need be considered. When extracting a short between a pin and a neighbor net, it may so happen that there are other geometries lying between the pin and the net. In this case, the extracted short should involve all the intersected geometries and not just the pin and the net. However, it is not possible to do that if the visibility criterion is ignored. One solution to this problem is to perform geometric intersections between the layout geometry that represents the short and the other layout geometries. This is the same as extracting multiple faults directly from the layout, which, as has been said, leads to a computationally expensive problem. As will be seen in Section 4.4.2, ignoring the visibility criterion only affects the diagnostic capability of the schemes based on graph coloring. The methods based on color mixing that will be explained in Section 4.4.3 and Section 4.4.4 are not significantly affected by the absence of the visibility criterion. Therefore, provided that color mixing techniques are used, ignoring the visibility criterion in the pin-to-net mode poses no major problem.

4.1.3 Implementation

The fault extraction procedure associates an open fault o_i to each net n_i, and extracts a short $e_{i,j}$ for any two nets whose distance is less than R. The complete fault extraction procedure is given by Algorithm 4.1 below.

In the pin-to-pin extraction mode, the complexity of Algorithm 4.1 is $O(N_P^2)$, where N_P is the number of pins. In the pin-to-net mode the complexity of the algorithm is $O(N_P \times N_S)$, where N_S is the number of geometries in the EA .[3] It is reasonable to assume that the ratios N_P/N and N_S/N do not increase with the circuit size. In fact, these ratios depend on the layout design style more than on the layout size. Hence, in simpler terms, it can be said that Algorithm 4.1 has complexity $O(N^2)$. In the limit, fault extraction can be performed in time $O(N \log N)$, provided the layout geometries are pre-sorted according to their geometric coordinates. It can be shown that in this case the complexity of searching for neighbor nets becomes $O(\log N)$. Since $O(N \log N)$ is the average complexity of common sorting algorithms (Sedgewick, 1983), the overall complexity of fault extraction becomes $O(N \log N)$.

Algorithm 4.1 has been integrated in `Pcb`, a public domain software package for PCB design (Nau, 1995a; Nau, 1995b). The `Pcb` source code is written in the `C` language (Kernighan and Ritchie, 1978), with a `X-Windows/Xt` graphical user interface (Jones, 1989; Young, 1990). The existing data structure has been in large part reused, with only a few modifications added to it. In this

---

[2]It is being assumed that all the ICs in the EA have the same pin spacing. This is true if, for example, all the ICs in the EA have SMT packages. In a more general case, a mix of IC packages may be present, and therefore variations in pin spacings must be accounted for.

[3]Note that each net is formed by a set of electrically connected geometries.

way, the implementation of Algorithm 4.1 has been greatly simplified (Sousa et al., 1996c). Also, it is natural that design, test and diagnosis tools are all integrated in the same environment, because most data structures can be shared among different tools. Most commercial software packages for Electronic Design Automation (EDA) integrate a myriad of tools in a common framework.

Figure 4.1 depicts the main window of `Pcb`, showing a PCB layout and the extracted faults. The opens are represented by component pins themselves; the shorts, extracted in the pin-to-net mode, are represented by the thin lines starting at a pin and ending at the closest point of a neighbor net. A dummy interconnection layer named `shorts` is used to place the shorts. The geometries on this layer can be manipulated like any other layout geometries. Shorts can be made visible/invisible, selected/unselected, deleted, etc. All these functional features can be useful if the package is extended to integrate tools for fault isolation.

Algorithm 4.1 (extractFaults(layout, mode, R)) :

```
O = ∅;   /* initialize the set of opens */
E = ∅;   /* initialize the set of shorts */
for each /*net*/ n_i do
    o_i = ∅;   /* initialize the net open */
    for each /*pin*/ P ∈ n_i do
        o_i = o_i ∪ P;   /* add a new possible location */
        for each /*net*/ n_j != n_i do
            e_i,j = ∅;   /* initialize the short */
            if(mode == PIN_TO_PIN)
                for each /*pin*/ Q ∈ n_j : PQ‾ < R do
                    e_i,j = e_i,j ∪ PQ;   /* add a new possible location */
            if(mode == PIN_TO_NET)
                for each /*point*/ Q ∈ n_j : PQ‾ < R do
                    e_i,j = e_i,j ∪ PQ;   /* add a new possible location */
            if(e_i,j != ∅)
                E = E ∪ e_i,j;   /* add to set of shorts */
    O = O ∪ o_i;   /* add to set of opens */
return O ∪ E;   /* produces the full set of single faults */
```

The input of the fault extraction tool is a layout file described in the `Pcb` layout format. A summary of the syntax of this file and an example file are given in Appendix A. With the output of fault extraction, a *graph* called the *adjacency graph* can be constructed (Garey et al., 1976). A *graph* is a mathematical abstraction consisting of a set of dots called *vertices* linked by a set of lines called *edges*. Graphs are useful mathematical tools and a rich set of graph algorithms can be found in the literature (McHugh, 1990). A graph G is usually denoted as $G = (C, E)$, where C is the set of vertices and E is the set of edges. The adjacency graph represents the extracted faults and is defined as follows:

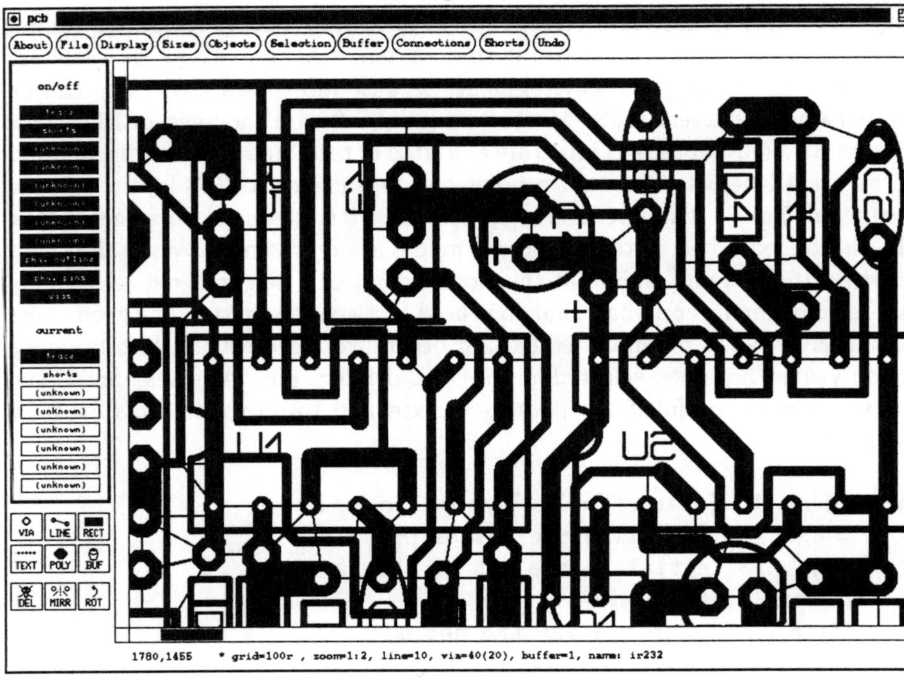

Figure 4.1. Pcb main window.

Definition 4.2 (Adjacency graph) *The adjacency graph is a graph $G = (C, E)$, where each vertex in C represents a net n_i, or an open fault o_i, and each edge in E represents a single short $e_{i,j}$ extracted from the layout.*

Graph G can be stored in two formats: (1) as an incidence matrix $A(G)$, where $A_{i,j}(G) = 1$ if $e_{i,j} \in E$, and 0 otherwise; or (2) as a linked list $L(G) = (l_1, l_2, \dots, l_N)$, where each element l_i is itself a linked list containing the neighbors n_j of each net n_i.

Example 4.1 *Consider the PCB layout shown in Figure 4.2(a). Applying Algorithm 4.1 to this layout, the adjacency graph G shown in Figure 4.2(b) is obtained. The pin-to-net extraction mode has been used, for a radius R as given in Figure 4.2(a).*

4.2 Diagnostic analysis

In this section a structural diagnostic analysis method is examined. The method produces layout level diagnostic information using the structural information provided by graph G.

In order to study the structural diagnostic analysis method, the concepts of set of suspect nets (Definition 3.2), partial diagnosis (Definition 3.3) and

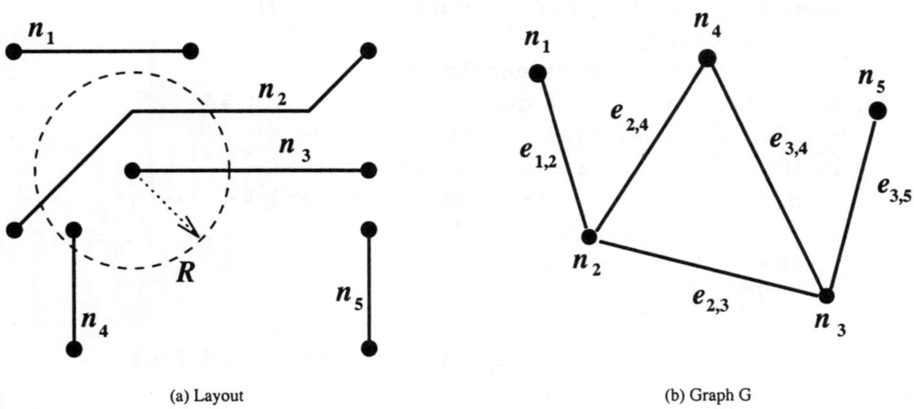

Figure 4.2. Graph G extraction criterion.

diagnosis (Definition 3.4), given in the context of behavioral diagnosis, will be replaced, respectively, by the following:

Definition 4.3 (Suspect subgraph (stain)) *A suspect subgraph, or stain, is a connected subgraph $T_j = (U,\ E')$ of G, whose nets $n_i \in U$ respond with the same vector $SRV_i = SRV_U$, which is erroneous for at least one net in U.*

Definition 4.4 (Partial diagnosis) *Given a stain $T_j = (U,\ E')$, the partial diagnosis F'_j associated with T_j is the set of all single faults in T_j. These are all single opens $o_i \in U$, and all single shorts $e_{i,k} \in E'$.*

Definition 4.5 (Diagnosis) *A diagnosis F' for a fault F is the union graph of all partial diagnoses F'_j of F.*

Algorithm 4.2 outlined below is a possible structural diagnostic analysis procedure. It is similar to the behavioral diagnostic analysis procedure of Algorithm 3.1. The difference between the two algorithms is that Algorithm 3.1 uses sets of suspect nets, whereas Algorithm 4.2 uses more detailed structural information given by stains. Like Algorithm 3.1, diagnostic analysis is performed by applying the analytic fault models S-R, A-R and S-A-0/1, studied in Chapter 2, in order to identify the partial diagnoses F'_j that will form a complete diagnosis F'.

The complexity of Algorithm 4.2 is given by the complexity of the stain identification step (line 1). This step is implemented using the linked list representation $L(G)$ of graph G. Each net n_i is searched for an erroneous response SRV_i. When an erroneous response SRV_i is found, the list $l_i \in L(G)$ is searched to check if the neighbors of n_i have the same erroneous response SRV_i. Each net checked is removed from $L(G)$, so that it is considered only once. This procedure has linear time complexity $O(N)$.

Algorithm 4.2 (analyzeGraphResponse(G, MRV)) :

ST = *getSetOfStains*(G, MRV);
$F' = \emptyset$; `/* initialize diagnosis */`
for each `/*stain*/` $T_j \in ST$ **do** { `/*` $T_j = (U, E')$ `*/`
    $F'_j = \emptyset$; `/* initialize partial diagnosis */`
    if($\#U == 1$) `/* a single suspect net */`
        if($SRV_U == \underline{0}/\underline{1}$) `/* apply S-A-0/1 and A-R */`
            $F'_j = \{e_{i,G/P},\ o_i : n_i \in U\}$;
        else `/* apply A-R only */`
            $F'_j = \{o_i : n_i \in U\}$;
    if($\#U > 1$) `/* multiple suspect nets */`
        if($SRV_U == \underline{0}/\underline{1}$) `/*apply S-R, S-A-0/1 and A-R*/`
            $F'_j = \{e_{i,k},\ e_{i,G/P},\ o_i : e_{i,k} \in E',\ n_i \in U\}$;
        else `/* apply S-R and A-R */`
            $F'_j = \{e_{i,k},\ o_i : e_{i,k} \in E',\ n_i \in U\}$; }
    $F' = F' \cup F'_j$;
return F';

The classification of diagnoses presented in Section 3.1 also applies to the structural diagnoses obtained with Algorithm 4.2. Whence, structural diagnoses can be complete, incomplete or incorrect; complete or incomplete diagnoses can be ambiguous, unambiguous or perfect. Despite the similarities between Algorithm 4.2 and Algorithm 3.1, the kind of diagnostic information provided by the two algorithms fundamentally differs. Algorithm 3.1 identifies shorted nets and open nets, but does not give any information about their physical location. On the other hand, Algorithm 4.2 can identify defect locations using its more elaborate data structure. A diagnostic strategy that combines Algorithm 4.2 and a good diagnostic synthesis scheme can be powerful. However, if the diagnostic synthesis scheme is poor, it may be better to simply use an entirely behavioral strategy.

Example 4.2 *Suppose that for a circuit with a fault F, both behavioral and structural diagnoses are performed. Behavioral diagnostic analysis is performed with Algorithm 3.1, and structural diagnostic analysis is performed with Algorithm 4.2.*

Suppose Algorithm 3.1 produces the behavioral diagnosis $F_b = \{e_{i,j}\}$, and Algorithm 4.2 produces the structural diagnosis $F_s = \{e_{i,j},\ e_{j,k}\}$.

To find the defects potentially identified by diagnosis F_b, nets n_i and n_j must be carefully inspected, since it is not known which physical defects can cause short $e_{i,j}$. Suppose, this operation takes time T_b. To find the defects potentially identified by diagnosis F_s, it is necessary to inspect the defect locations previously extracted for shorts $e_{i,j}$ and $e_{j,k}$. Suppose, this operation takes time T_s.

If $T_s < T_b$ than the structural approach is clearly superior to the behavioral approach. However, since F_s identifies two shorts and F_b identifies only one, it may take more time to inspect all defects associated with F_s than to inspect the *two nets associated with F_b. In this case $T_s > T_b$, and the structural approach becomes inferior to the behavioral approach.*

Now suppose that a more powerful diagnostic synthesis scheme is used, and that the structural approach produces the less ambiguous diagnosis $F'_s = \{e_{i,j}\}$. The time taken to find the respective defects is T'_s, which, of course, is shorter than the time T_s taken by the more ambiguous diagnosis F_s. Since both F_b and F'_s contain one short, it is quite likely that T_s is significantly shorter than T_b; inspecting the pre-extracted defects associated with one short is normally faster than inspecting two entire nets for shorts between them.

Therefore, structural diagnostic analysis is worth the effort only if a good diagnostic synthesis scheme is employed. Otherwise, it may be better and cheaper to simply use behavioral diagnosis.

4.3 Fault isolation

Algorithm 3.2, the fault isolation procedure in behavioral diagnosis, also applies to structural diagnosis. However, fault isolation is greatly aided by the fact that all faults are mapped to physical defects. The function *findDefects()* in Algorithm 3.2 can be much more efficient, since the locations of the potential defects of the fault passed as argument have been previously stored. Function *findDefects()* only needs to inspect those locations to check if the defects have occurred.

Automatic repair technology will need good fault isolation algorithms, able to guide repair robots to the defect locations. This is only possible if structural information is available. Automatic repair technology is therefore enabled by structural information. One of the main thrusts of the present methods is their usefulness in future developments of automatic repair techniques, with all the economic benefits that can be brought by such technology. The structural perspective in interconnect diagnosis introduced in this work is a good step towards that direction.

It may be argued that automatic repair technology can be developed independently from diagnosis, which, then, could be purely behavioral. That would correspond to automating the behavioral fault isolation procedure given by Algorithm 3.2. The main difficulty would be in implementing the function *findDefects()* in Algorithm 3.2, which is equivalent to implementing a fault extraction procedure *a posteriori*. It would run after potential faults are diagnosed, delaying the use of structural information to a later development stage. The early fault extraction approach proposed in this work seems more sensible. In fact, as will be seen in the next section, structural information is also useful for generating compact test sequences. If the extraction of physical information is delayed to a later stage, this oportunity is completely missed.

4.4 Diagnostic synthesis

In structural diagnostic synthesis the goal is the same as in behavioral diagnostic synthesis — minimize STV length and maximize diagnostic capability.

Also, the same synthetic fault models of Section 2.2 are used: the wired-logic (W-A/O), strong-driver-logic (S-D-k) and stuck-at-logic (S-A-0/1) models. In this section the existing methods and a new method are described and analyzed.

As discussed in Section 3.3, the minimum requirement of a diagnostic synthesis scheme is the detectability requirement. In behavioral diagnostic synthesis, a necessary and sufficient condition for detectability has been identified in Theorem 3.1. In structural diagnostic synthesis, an equally general result could not be obtained. However, it is possible to show that for *restricted topology faults* detectability can be proved, which suffices in most practical situations.

Definition 4.6 (Restricted topology fault) *A fault is said to have a restricted topology if the nets it affects are at least connected to a driver or to a receiver; in other words, the fault must not cause floating nets.*

According to Definition 4.6, non-restricted topology faults are faults that cause floating nets. Intuitively, these faults should be very unlikely. Note that, for a net with a single driver and a single receiver, at least two open defects are needed to create a floating net. The theoretical results in this chapter are only valid for restricted topology faults. The first result is the condition for detectability, which is established by the following theorem:

Theorem 4.1 *According to the synthetic fault models, a necessary and sufficient condition for detectability of any restricted topology fault F is that all the STVs in the test set are different from $\underline{0}$ or $\underline{1}$, and that neighbor nets have different STVs.*

Proof: *Similar to the proof of Theorem 3.1.* ■

It will now be explained why Theorem 4.1 can only be proved for restricted topology faults, while Theorem 3.1 is proved for any topology faults. In the proof of Theorem 3.1, it is shown that, in order to justify open net receivers, new faults have to be added in an infinite recursion. This is a contradiction, since the number of faults in a circuit must be finite, which proves the impossibility of undetectable faults. Similarly, in the proof of Theorem 4.1 new faults have to be added in an infinite recursion in order to justify open net receivers. The restricted topology constraint is needed to ensure that no floating net segments can bridge two nets assigned the same STV. Note that, unlike in behavioral diagnosis, where each net has a unique STV, in structural diagnosis two nets may have the same STV if they are not direct neighbors in the layout. Therefore, if a floating net bridges two nets originally assigned the same STV, an undetectable fault may be produced if the floating net itself also eludes detection. That is why floating nets are not allowed, and therefore the result is only valid for restricted topology faults.

With the concept of detectability in place, the following definition of a structural diagnostic scheme can be given:

Definition 4.7 (Structural interconnect diagnostic scheme) *A test set MTV is an interconnect diagnostic scheme if it verifies the minimum requirement for detectability imposed by Theorem 4.1.*

Any structural interconnect diagnostic scheme has the property of ensuring diagnosability. The following theorem establishes this result:

Theorem 4.2 *According to the synthetic fault models, any restricted topology fault F will be diagnosed in a finite number of test-diagnose-repair iterations, if a diagnostic scheme is repeatedly applied, followed by the diagnostic analysis procedure given by Algorithm 4.2 and the fault isolation procedure given by Algorithm 3.2.*

Proof: *Similar to the proof of Theorem 3.2.* ∎

Theorem 4.2 establishes that a structural diagnostic scheme is complete for the synthetic fault models. Now, the capability of structural diagnostic schemes is studied in terms of the aliasing and confounding concepts. The concepts of set of shorted nets and faulty subset, given respectively by Definition 2.2 of Section 2.2.3 and Definition 3.7 of Section 3.3, are replaced by the following ones:

Definition 4.8 (Subgraph of shorted nets) *A subgraph $S = (V, E'')$ of shorted nets is a connected subgraph, in which the nets $n_i \in V$ produce the same response $SRV_i = SRV_S$, and the edges $e_{i,j} \in E''$ represent single short faults between nets $n_i, n_j \in V$.*

Definition 4.9 (Faulty subgraph) *A faulty subgraph $H = (V, E'')$ is a connected subgraph of a stain $T = (U, E')$, whose response $SRV_V = SRV_U$ can be explained in terms of some synthetic fault model involving the nets in V only.* [4]

The concepts of aliasing and confounding, given in the context of behavioral diagnosis by Definitions 3.6 and 3.8, respectively, both from Section 3.3, also need be replaced by the following ones:

Definition 4.10 (Structural aliasing) *A structural diagnostic scheme is said to be aliasing with respect to some synthetic fault behavior if that behavior can produce a stain $T = (U, E')$, such that U includes at least a fault-free (aliased) net n_i.*

Definition 4.11 (Structural confounding) *A structural diagnostic scheme is said to be confounding with respect to a pair of synthetic fault behaviors if those fault behaviors can create a stain $T = (U, E')$, which can be explained, by a pair of disjoint faulty subgraphs H_1 and H_2 of T.*

Structural aliasing and confounding are similar to behavioral aliasing and confounding. However, they describe ambiguity in local rather than in global

[4] The above two definitions should not be confused: a subgraph of shorted nets is the graph representation of a short fault; a faulty subgraph is an attempt to explain the observed response in terms of some synthetic fault model.

terms. In other words, the ambiguity produced by a structural scheme is always contained within the boundaries of a stain. Structural aliasing and confounding can be used to compare a structural scheme to another structural scheme. In the same way, behavioral aliasing and confounding can be used to compare a behavioral scheme to another behavioral scheme. However, it is difficult to compare a behavioral to a structural scheme in terms of these concepts. For example, the diagnostic capability of an aliasing-free behavioral scheme may be totally different from the diagnostic capability of an aliasing-free structural scheme. This issue will be treated in detail in Chapter 5. For the time being, it suffices to say that precise comparisons between behavioral and structural schemes can only be made with a quantitative diagnostic resolution framework. The qualitative aliasing/confounding approach is not adequate for this purpose.

In the following subsections four different methods are presented. In Section 4.4.1, an obvious but original method is discussed: structural diagnostic analysis using behavioral vectors. In Section 4.4.2, a method based on *graph coloring* (Cheng et al., 1990) is critically reviewed in the light of the concepts introduced in this work. In Section 4.4.3, another known method based on *color mixing* is discused. Last, in Section 4.4.4, the structural diagnosis method proposed in the scope of this research, the *statistical color mixing* method, is described.

4.4.1 Behavioral vectors

A straightforward way to perform structural interconnect diagnosis is by restricting the use of structural information to diagnostic analysis. The test set is generated using one of the schemes from Section 3.3, and diagnostic analysis is performed using Algorithm 4.2.

With behavioral diagnostic analysis, if a single set of suspect nets U occurs, the number of possible single shorts in a diagnosis F' is $\binom{\#U}{2}$. Now suppose that structural diagnostic analysis is being used, and that U produces a stain $T = (U, E')$. Then, the number of possible single shorts in F' is $\#E'$. Unless T is a *fully connected graph*, [5] which it rarely is, then in most cases $\#E' < \binom{\#U}{2}$, implying a lower diagnostic ambiguity than with behavioral diagnostic analysis. Furthermore, a fault that appears as a single set of suspect nets U in behavioral diagnosis may appear as multiple unconnected stains in structural diagnosis. In this case, the diagnostic ambiguity is even lower; in general, the number of edges in multiple unconnected stains is lower than the number of edges in a single stain containing the same number of vertices.

Overall, with behavioral vectors, structural aliasing and structural confounding responses become intrisically unlikely. A net must be part of a stain in order to be aliased; faulty subgraphs also must form the same stain in order to be confounded. These constraints make aliasing and confounding much less probable. The main advantage of behavioral diagnosis — test set layout in-

[5] A fully connected graph is a graph where all pairs of vertices are connected by edges.

dependence — is retained, while adding the benefits of structural diagnostic analysis. Since graph G is only needed for diagnostic analysis, its extraction can be postponed to a later development stage, when there is little chance of further layout modifications.

Despite the convenience of structural diagnosis with behavioral vectors, a great opportunity is being missed: structural information can be used not only to facilitate diagnostic analysis but also to reduce the length of the test vector sequences. In fact, each net can only short to its neighbor nets in the layout, and not to any other net. This piece of information greatly reduces the number of possible faults that need to be considered, enabling significant reduction in the length of the STVs. In the next section, a scheme that oportunistically exploits this possibility is described.

4.4.2 Graph coloring vectors

Graph coloring techniques can be used to identify sets of nets that cannot short directly to each other. These nets can be assigned the same STV, and, because less unique STVs are required, the STVs can be shorter. This technique was originally proposed to generate test vectors for detecting shorts in bare PCBs (Garey et al., 1976).

A graph $G = (C, E)$ is said to be k-colorable if k colors can be used to color the vertices of G, so that neighbor vertices are always assigned different colors. If C is a set of vertices and CS is a set of colors, the graph coloring problem is that of finding a color mapping $CM(C) : C \rightarrow CS$, so that the total number of colors k is minimized. Such minimum value k_{min} is referred to as the *chromatic number* of the graph. The graph coloring problem is a well known NP-complete problem (Garey and Johnson, 1979).

Graph coloring vectors are obtained by coloring the adjacency graph G and then assigning a unique STV to each set of nets that have the same color. Despite being an NP complete problem, it has been shown that adjacency graphs can be colored with a small number of colors $k \ll N$, using fast polynomial complexity heuristics (Garey et al., 1976). Therefore, few unique STVs are needed, as large numbers of nets will have the same color. If few STVs are required then their length can be significantly lower when compared to behavioral diagnostic synthesis. The fact that adjacency graphs can be colored with a few colors stems from the fact that adjacency graphs are obtained from circuit layouts, which give rise to *planar* or *nearly planar* graphs.

Definition 4.12 (Planar graph) *A* planar graph *is one that can be laid out on the plane without edges intersecting one another.*

It can be shown that a graph extracted from a single layout layer is planar. Planar graphs can be colored with no more than 4 colors, as established by the famous *four-color theorem* (Appel and Haken, 1977).

Definition 4.13 (Nearly planar graph) *A* nearly planar *graph is one that is extracted from a layout containing two layers.*

Adjacency graphs derived from PCBs are nearly planar because they are extracted from at most two layers, which correspond to the two sides of a board. Intermediate layers are ignored because they are not exposed to defects that occur during the assembly process. The bare board is assumed to be pretested and defect-free, so that the internal layers will remain intact. By analogy with planar graphs, it is natural to expect a low chromatic number for nearly planar graphs. However, an upper bound for the chromatic number of nearly planar graphs is unknown.

In the short detection problem, it suffices that colors are coded with the minimum number of bits (Garey et al., 1976). This corresponds to a counting sequence scheme (Section 3.1) for coding colors instead of nets. It has been suggested that the approach can be extended to the short diagnosis problem by simply using more elaborate color coding schemes (Cheng et al., 1990). According to this suggestion, if an aliasing-free scheme is desired, then the colors should be coded with an aliasing-free scheme like, say, the counting sequence plus complement scheme. This idea is the basis for the following algorithm:

Algorithm 4.3 (generateGraphColorVecs(*G*, *coding_scheme*)) :

1. *CM(C) = colorGraph(G);* `/* see Appendix B */`
2. *MTV = codeColors(k, coding_scheme);* `/* codes the` *k* `colors using` *coding_scheme* `*/`
3. **return** *MTV;*

The complexity of Algorithm 4.3 is given by the complexity of the graph coloring step (line 1). A polynomial time graph coloring heuristic, Algorithm B.1, is described in Appendix B. Algorithm B.1 is used in the experiments reported in Chapter 6, and has complexity $O(N^3)$. Thence, Algorithm 4.3 has time complexity $O(N^3)$. In Appendix B, it is also shown how to improve Algorithm B.1 with a $O(D \times N)$ graph coloring heuristic, D being the average vertex degree .[6]

The graph coloring approach represents a breakthrough in solving the short detection problem. However, the suggestion by Cheng et al. to extend the method to interconnect fault diagnosis does not, unfortunately, work as expected. Contrary to that suggestion, the aliasing and confounding properties of a graph coloring scheme do not match the properties of the corresponding behavioral scheme. This will be shown using three schemes: the modified counting sequence, the self-diagnosis sequence, and the K sequence (which is the equivalent of the N sequence when it is applied to k colors rather than to N nets). The remainder of this section will discuss the application of these three sequences to color coding.

Modified counting sequence color coding. In behavioral diagnosis the STVs can be viewed as codes assigned to the nets. Each net must have a unique STV to qualify as a diagnostic scheme, which means each net must

---

[6] The degree of a vertex n_i in a graph G is the number of edges $e_{i,j}$ of vertex n_i.

have a distinct code. In structural diagnosis by graph coloring, it is each color (and not each net) that needs a distinct code. Therefore, the STV length of a graph coloring scheme can be obtained by simply replacing N with k in the expression that gives the STV length of the corresponding behavioral scheme. For the modified counting sequence, replacing N with k in Equation (3.2) yields

$$p = \lceil \log_2(k+2) \rceil. \tag{4.1}$$

Due to the fact that $k \ll N$, the number of unique STVs required by the modified counting sequence color coding scheme is drastically lower than the number of unique STVs required by the behavioral modified counting sequence scheme. As a result, the STV length is also reduced, but the reduction is not as spectacular as the reduction from nets to colors. According to Equation (3.3) and Equation (4.1), the STV length is a logarithmic function of the number of nets in the behavioral scheme, and a logarithmic function of the number of colors in the structural scheme, respectively. Nevertheless, the reduction in STV length is still significant as illustrated by the following example:

Example 4.3 *Suppose an EA has $N = 2^{16}$, i.e., 65536 nets, and its graph G is 8-colorable (this is a realistic figure). The modified counting sequence in behavioral diagnosis needs an STV length $p = \lceil \log_2(2^{16}+2) \rceil = 17$. On the other hand, in the graph coloring scheme the STV length is $p = \lceil \log_2(8+2) \rceil = 4$. That is, the graph coloring test set is about four times shorter than the behavioral test set.*

To illustrate the modified counting sequence color coding scheme, consider the graph G shown in Figure 4.2(b). After the graph coloring step, the 3-color graph shown in Figure 4.3 is obtained. Then the colors are coded and the STV mapping obtained is given in Table 4.1.

Table 4.1. Modified counting sequence color coding for graph G.

Color	Nets	STV
c_1	n_2	01
c_2	n_1, n_4, n_5	10
c_3	n_3	11

Because the modified counting sequence fulfills the minimum requirement for detectability imposed by Theorem 4.1, it qualifies as a diagnostic scheme according to Definition 4.7.

The following lemma (which is a version of Lemma 3.2 adapted to structural diagnosis) can be proved for any structural diagnostic scheme:

Lemma 4.1 *Any structural diagnostic scheme is aliasing-free for (a) S-A-0/1 and (b) S-D-k faults.*

Proof: *Similar to the proof of Lemma 3.2* ■

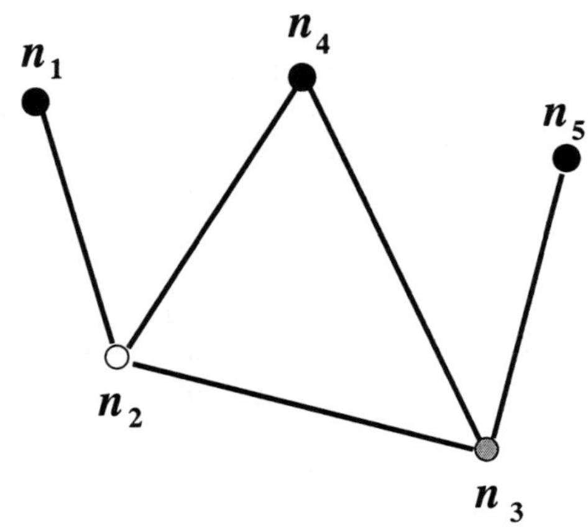

Figure 4.3. A 3-color mapping for graph G of Figure 4.2(b).

The above lemma shows that the structural aliasing capability of the modified counting sequence graph coloring scheme corresponds to the behavioral aliasing capability of the behavioral modified counting sequence scheme. However, it will now be shown that the confounding capabilities do not match. The following lemma (which is a version of Lemma 3.3 for structural diagnosis) establishes some confounding capabilities valid for any graph coloring scheme:

Lemma 4.2 *In any structural diagnostic scheme, any pair of faulty subgraphs H_1/H_2 cannot confound if characterized by the pairs of behaviors S-A-0/S-A-1, S-A-0/W-O, S-A-1/W-A, S-A-0/S-D-k and S-A-1/S-D-k.*

Proof: *Note that $H(SRV_{V_1}) \neq H(SRV_{V_2}) \Rightarrow SRV_{V_1} \neq SRV_{V_2}$.* ∎

By comparing Lemma 4.2 to Lemma 3.3, it is noticeable that, while in a behavioral scheme S-D-k/S-D-l confounding cannot occur, in a structural scheme it may occur. This is because net n_k, the strong-driver of H_1, is not necessarily assigned a different STV from n_l, the strong-driver of H_2. Only in the case where n_k and n_l are neighbors, will they have different STVs. This is the first example of the differences between the diagnostic properties of behavioral and structural schemes. However, this difference is not yet very significant. As a matter of fact, for this kind of confounding ambiguity to occur, it is necessary that the subgraphs H_1 and H_2 are *adjacent* in G:

Definition 4.14 (Adjacent subgraphs) *Two subgraphs are said to be adjacent if they are separated by exactly one edge in the graph they belong to.*

If faults are seen as random spatial events, the probability that two subgraphs of shorted nets will be adjacent can be deemed to be very low, even for a modest size layout. Multiple faults are more likely to form a connected subgraph, as they normally result from the same physical cause. For example, a large solder blob may accidentally short 3 nets, causing a 2-edge subgraph in G. It is true that, if adjacent faults occur the chances that confounding will happen become quite high, since there are few unique STVs in a graph coloring test set. However, because adjacent faults are unlikely, the overall probability of confounding should not be significant.

More significant differences between structural and behavioral diagnostic properties will be found in the coding schemes. Such differences cannot be dismissed as improbable and pose real difficulties to structural diagnostic synthesis. As for the modified counting sequence color coding scheme, its discussion will be concluded after the following theorem establishes its diagnostic capability:

Theorem 4.3 *A test set MTV obtained by Algorithm 4.3, using the modified counting sequence color coding scheme, is (a) aliasing-free for S-A-0/1 faults and (b) confounding-free for the pairs of behaviors S-A-0/S-A-1, S-A-0/W-O, S-A-1/W-A, S-A-0/S-D-k and S-A-1/S-D-k.*

Proof: *Clause (a) and clause (b) follow from Lemma 4.1 and Lemma 4.2, respectively.* ■

<u>**Self-diagnosis color coding.**</u> The self-diagnosis graph coloring scheme is obtained by coding the colors with the self-diagnosis sequence studied in Section 3.3.5. The STV length is obtained by replacing N with k in Equation (3.8):

$$p = \min j, \qquad \text{such that} \qquad \binom{j}{\lfloor \frac{j}{2} \rfloor} \geq k. \tag{4.2}$$

Inequality (3.9) has established that the behavioral self-diagnosis STV length has logarithmic complexity in the number of nets N. Accordingly, the self-diagnosis graph coloring STV length has logarithmic complexity in the number of colors k. As shown for the modified counting sequence, the STV length of the structural self-diagnosis scheme is significantly lower than the STV length of the behavioral self-diagnosis scheme, and increases slowly with the number of colors.

As for the diagnostic capability, the following lemmas and theorem will state the properties of the self-diagnosis graph coloring scheme:

Lemma 4.3 *A test set obtained with Algorithm 4.3, using independent STVs for color coding, is aliasing-free for (a) S-A-0/1 and (b) W-A/O faults.*

Proof: *Clause (a) follows from Lemma 4.1, since the STVs $\underline{0}/\underline{1}$ cannot be present in a set of independent STVs. For clause (b) let the W-A case be proved; the W-O case can be proved by duality. A W-A short S contains at least one edge $e_{i,j}$ between two neighbor nets n_i and n_j. Since STV_i and STV_j are independent, $SRV_S \subset STV_i, STV_j$. Thus, it must be true that $SRV_S \notin MTV$, which implies that W-A aliasing is impossible.* ■

Lemma 4.4 *In a test set obtained with Algorithm 4.3, using independent and fixed weight STVs of weight W for color coding, any pair of disjoint faulty* *subgraphs H_1/H_2 cannot confound if they are characterized by a pair of different behaviors, except for the S-A-0/W-A, S-A-1/W-O and S-D-k/S-D-l pairs.*

Proof: *To prove this lemma one needs to prove the cases S-A-0/S-A-1, S-A-0/W-O, S-A-0/S-D-k, S-A-1/W-A, S-A-1/S-D-k, W-A/W-O, W-A/S-D-k and W-O/S-D-k. For all these cases $H(SRV_{V_1}) \neq H(SRV_{V_2})$, which means that $SRV_{V_1} \neq SRV_{V_2}$.* ∎

Theorem 4.4 *A test set obtained with Algorithm 4.3, using a self-diagnosis sequence for color coding, is (a) aliasing-free for S-A-0/1 and W-A/O behaviors, and (b) confounding-free for any pair of different behaviors, except for the S-A-0/W-A, S-A-1/W-O and S-D-k/S-D-l pairs.*

Proof: *Clause (a) follows from Lemma 4.3, since the self-diagnosis sequence is a set of independent STVs by definition. Clause (c) follows from Lemma 4.4, since the self-diagnosis sequence by definition is a set of fixed weight STVs of weight $W = \lfloor p/2 \rfloor$.* ∎

Now, a more significant difference between behavioral and graph coloring schemes becomes evident: whereas the behavioral self-diagnosis scheme is aliasing-free, the self-diagnosis graph coloring scheme is not (Sousa et al., 1996b). In fact, the following example shows how the self-diagnosis graph coloring scheme is aliasing for S-D-k faults:

Example 4.4 *Suppose a net n_i responds with a correct response $SRV_i = STV_i$, and is a neighbor to a short S with behavior S-D-k. If it so happens that n_i and n_k have the same color, that is, $STV_i = STV_k$, then net n_i is aliased according to Definition 4.10.*

The S-D-k aliasing problem is due to the fact that a stain may contain nets originally assigned the same color. Unlike the S-D-k/S-D-l confounding problem studied above, the probability of S-D-k aliasing cannot be neglected. Since graph coloring minimizes the number of colors used, it is quite likely that an S-D-k short will alias one of its neighbor nets. Worst of all, the root of the problem is the non-uniqueness of the STVs, the very essence of structural diagnostic synthesis. The problem is unsolvable whenever STVs are reused for different nets. Yet, as will be seen, it is possible to reduce the probability of such occurrences, if a small increase in SVT length is tolerated.

K sequence color coding. The graph coloring counterpart of the N sequence (Section 3.3.8) has been called the K sequence, because its STV length is obtained by replacing N with k in Equation (3.15):

$$p = k. \tag{4.3}$$

The STV length in the K sequence is a linear funtion of the number of colors k, inasmuch the same way the STV length in the N sequence is a linear function of the number of nets N. Therefore, if the structural K sequence is used instead of the behavioral N sequence, the STV length reduction is as drastic as the reduction from N nets to k colors. Recall that the same does not happen with the modified counting sequence or the self-diagnosis sequence, which have a logarithmic dependence on the number of colors.

Since the diagnostic capability of the N sequence is good, according to the suggestion by Cheng et al., one is led to expect a similarly good capability for the K sequence. But, unfortunately, the drastic reduction in STV length is accompanied by an equally drastic reduction in the diagnostic capability. This can be better understood after the diagnostic capability of the K sequence is established by the following theorem:

Theorem 4.5 *A test set obtained with Algorithm 4.3, using the K sequence for color coding, is (a) aliasing-free for S-A-0/1 faults, and (b) confounding-free for the behavior pairs S-A-0/S-A-1, S-A-0/W-O, S-A-1/W-A, S-A-0/S-D-k and S-A-1/S-D-k.*

Proof: *Clause (a) follows from Lemma 4.1, and clause (b) follows from Lemma 4.2.* ∎

Theorem 4.5 clearly shows how inferior the diagnostic capability of the graph coloring K sequence is when compared to the behavioral N sequence. Indeed, in the behavioral sense, the N sequence offers almost minimal diagnostic ambiguity (Theorem 3.9), whereas, in the structural sense, the K sequence is no better than the modified counting sequence graph coloring scheme (Theorem 4.3). That is, improving the coding scheme in structural diagnosis by graph coloring has little or no effect on the diagnostic capability. In contrast, in behavioral diagnosis the coding scheme is a decisive factor in terms of diagnostic capability.

The good diagnostic capability of the behavioral N sequence is mostly due to the global-diagnosability property (Section 3.3.6). In structural diagnosis by graph coloring, the global-diagnosability property does not apply, and that is why the K sequence diagnostic capability is so low. In fact, consider a stain T: if more than one net shows a correct response in T, it becomes ambiguous which of those nets are faulty. That is what happens with aliasing S-D-k shorts: both the strong-driver net and the aliased net produce correct responses. Diagnosis of strong-driver shorts always relies on the global-diagnosability property, since the strong-driver net is diagnosed by observing the response of the other nets shorted to it. The fact that global-diagnosability fails in structural diagnosis by graph coloring was not realized by Cheng et al. when they showed, incorrectly, that a W-O global-diagnosis color coding scheme is W-O aliasing-free (Cheng et al., 1990). A counterexample is given next:

Example 4.5 *For the graph in Figure 4.3, the W-O global-diagnosis graph coloring scheme is given in Table 4.2. If there is a W-O single short* $e_{1,2}$ *then a stain* $T = (U,\ E')$, *where* $U = \{n_1, n_2, n_4\}$ *and* $E' = \{e_{1,2}, e_{2,4}\}$, *appears on the graph, such that* $SRV_U = 10$. *Since* $SRV_4 = STV_4 = 10$, *according to Definition 4.10, net* n_4 *is aliased in* T.

Table 4.2. Structural global-diagnosis color coding for graph G.

Color	Nets	STV
c_1	n_1, n_4, n_5	10
c_2	n_2	00
c_3	n_3	01

The lack of global diagnosability in the graph coloring schemes also prevents the K sequence scheme from being confounding-free. According to Theorem 4.5, excluding the obvious cases involving S-A-0/1 faults, the K sequence cannot resolve any other confounding pairs. On the other hand, the corresponding behavioral N sequence can resolve almost all confounding pairs (Theorem 3.9). The explanation for this asymetry can be given in four steps: (1) most of the confounding immunity of the N sequence comes from the W-A and W-O diagonal independence property (Lemma 3.8); (2) in behavioral diagnosis, diagonal independence works well because it uses global diagnosis; (3) since it has been shown that global diagnosis only makes sense if the STVs are unique, it immediately follows that diagonal independence does not work in structural diagnosis by graph coloring. The confounding problem, like the aliasing problem, is unsolvable in any structural diagnostic scheme where STVs can be reused for different nets.

In conclusion, the diagnostic capability of a graph coloring scheme does not match the diagnostic capability of the corresponding behavioral scheme. The suggestion by Cheng et al., unfortunately, is inadequate. To improve the diagnostic capability of structural diagnostic synthesis, it is not enough to simply enhance the color coding scheme. As will be shown, procedures of a different kind are needed to achieve this goal.

4.4.3 Color mixing vectors

The key to improving structural diagnostic capability is to improve global diagnosability, mainly to eliminate structural aliasing caused by S-D-k shorts. This fact has not been realized by other researchers chiefly because S-D-k shorts were not addressed. Some researchers did notice the persistence of the W-A/O structural confounding problem, regardless of the color coding scheme being used. In terms of the occurrence probability, the S-D-k structural aliasing problem is more important than the W-A/O structural confounding problem. But, paradoxically, solving the latter has been the main driving force behind improving structural diagnostic schemes.

Attempts at solving this problem have been independently made by two research teams (Lien and Breuer, 1991; McBean and Moore, 1993). These attempts are fundamentally similar and are based on two principles: (1) adding extra edges to graph G to force the graph to be colored with more colors, and

(2) using a walking ones sequence to code the colors to resolve W-O confounding ambiguities. The procedure is outlined by Algorithm 4.4.

Algorithm 4.4 (generateColorMixVecs(G)) :
1. G_2=*enhanceGraph(G, 2);* `/* see Appendix C */`
2. $CM(C)$ = *colorGraph*(G_2); `/* see Appendix B */`
3. MTV = *codeColors(k, walking_ones);* `/* codes the k colors with the walking-1s sequence */`
4. **return** MTV;

The complexity of Algorithm 4.4 is given by the complexity of the graph enhancing step (line 1). The graph enhancing step is carried out by calling the function *enhanceGraph()* with parameters G and 2. The parameter G is obviously the graph to be enhanced. The second parameter is the *maximum stain extent*, a concept that will be explained in the next section. For the moment, it suffices to say that graph G_2 is obtained by adding an extra edge $e_{j,k}$ between any two vertices n_j and n_k in G that have a common neighbor n_i. The function *enhanceGraph()* has been implemented using Algorithm C.1, described in Appendix C. The complexity of Algorithm C.1 is $O(N^3)$ if the maximum stain extent is set to 2. In Appendix C, a more efficient algorithm, Algorithm C.2, is described, which, when the maximum stain extent is set to 2, has time complexity $O(N \times D^2)$.

The enhancing procedure prevents two single shorts from causing a W-A/O confounding response. Such a procedure is also known as the *triangulation transformation* (McBean and Moore, 1993), because the extra edge $e_{j,k}$ and the two original edges $e_{j,i}$ and $e_{i,k}$ form a triangle with vertices $\{n_i, n_j, n_k\}$. Figure 4.4 illustrates the triangulation transformation.

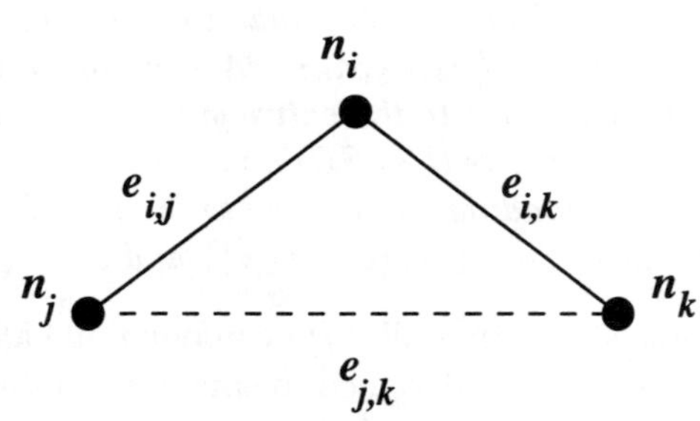

Figure 4.4. The triangulation transformation.

Unlike the original edges of G, the extra edges do not represent physical shorts. Their function is to force the addition of more colors, in order to reduce diagnostic ambiguity. Since G_2 has more edges than G, its average vertex degree D_2 will be greater than D, the average vertex degree of G. It can be shown that, in general, the chromatic number of a graph depends on the average degree of the graph. A higher average degree normally corresponds to a higher chromatic number. The designation *color mixing* comes from the fact that more colors can be more uniformly distributed locally in the graph. A lower bound for the chromatic number of G_2 can be established by the following lemma:

Lemma 4.5 *A lower bound for the chromatic number k_{min} of G_2, as obtained by Algorithm 4.4, is given by $k_{min} \geq D_{max} + 1$, where D_{max} is the maximum vertex degree in G.*

Proof: *A net n_i with degree D_{max} has D_{max} neighbors. After the graph enhancing step (line 2), n_i and its neighbors form a fully connected subgraph of G_2. After G_2 is colored (line 2), the $D_{max} + 1$ vertices in this fully connected subgraph all receive unique colors. Therefore, at least $D_{max}+1$ colors are needed to color G_2.* ■

As explained in Appendix B, due to the properties of adjacency graphs, common graph coloring algorithms produce a number of colors for G which is normally well below $D_{max}+1$. According to Lemma 4.5, the chromatic number of G_2 is normally well above the chromatic number of G. More colors mean longer STVs. The situation is aggravated by the fact that Algorithm 4.4 uses a walking sequence for color coding. Recall that, with a walking sequence, the STV length is linear in the number of colors.

The objective of Algorithm 4.4 is to achieve maximal diagnosis with respect to W-O shorts, [7] i.e., no aliasing and no confounding (Lien and Breuer, 1991; McBean and Moore, 1993). However, as concluded in the previous section, this goal cannot be achieved as long as there are nets assigned the same STV in the circuit. In fact, W-O confounding can still occur, as illustrated by the following counterexample:

Example 4.6 *Consider a linear graph G as given by Figure 4.5(a). Applying Algorithm 4.4, the 3-color graph G_2 of Figure 4.5(b), and the test set of Table 4.3, are obtained. Now suppose that two sets of shorted nets occur: $V_1 = \{n_1, n_2, n_3\}$ and $V_2 = \{n_4, n_5, n_6\}$. Assuming W-O behavior, a stain $T = (U, E')$, which is equivalent to the entire graph G, is produced, and the response of the set of suspect nets U is $SRV_U = 111$. Then, according to Definition 4.11, structural confounding occurs, due to the fact that T can be produced by the two disjoint shorts $S_1 = (V_1, \{e_{1,2}, e_{2,3}\})$ and $S_2 = (V_2, \{e_{4,5}, e_{5,6}\})$.*

The above example shows that W-A/O confounding can still occur, despite the fact Algorithm 4.4 has been designed to eliminate this ambiguity. Nonetheless, it can be argued that too many shorts are necessary to cause confounding, and, therefore, its likelihood can be neglected. In Example 4.6, four single shorts ($e_{1,2}$, $e_{2,3}$, $e_{4,5}$ and $e_{5,6}$) are necessary for W-O confounding to occur. This is a fair comment, as far as W-A/O behaviors are concerned. If strong-driver-logic behaviors are admitted, it takes just two shorts to cause confounding. The following example ilustrates this case:

[7] For W-A shorts, the walking-0s test set should be used instead of the walking-1s sequence. In fact both W-A and W-O shorts can be considered by using the complete walking sequence (Section 3.3.3), the N+1 sequence (Section 3.3.7) or the N sequence (Section 3.3.8).

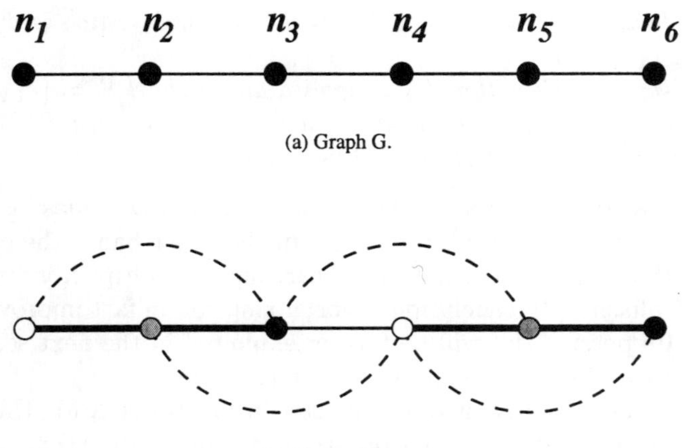

(a) Graph G.

(b) The 3-color graph G_2.

Figure 4.5. Color mixing counterexample.

Table 4.3. Test set for color mixing counterexample.

Color	Net	STV
c_1	n_1, n_4	100
c_2	n_2, n_5	010
c_3	n_3, n_6	001

Example 4.7 *Consider again the graph G in Figure 4.5(a), the 3-color mapping obtained with Algorithm 4.4, and the coding given in Table 4.3. Suppose the single shorts $e_{1,2}$ and $e_{3,4}$ occur with behaviors S-D-1 and S-D-4, respectively. In this situation, a confounding response is caused, since $STV_1 = STV_4$ implies $SRV_{1,2} = SRV_{3,4} = 100$.*

As explained in Section 4.4.2, it is unlikely that two adjacent shorts will occur. Multiple shorts forming connected subgraphs are more common, and tend to cause S-D-k aliasing, identified before as the main unsolved problem of structural diagnosis by graph coloring. As will be seen, Algorithm 4.4 does not completely solve this problem. Nevertheless, it unintentionally makes it less likely than with Algorithm 4.3. The following example clarifies this point:

Example 4.8 *Suppose Algorithm 4.3 is applied to graph G in Figure 4.5(a), and a 2-color mapping is obtained. The nets with an even subscript get a color c'_1, whereas the nets with an odd subscript get a color c'_2. If a short $S = \{e_{1,2}\}$ between n_1 and n_2 with behavior S-D-1 occurs, then net n_3 will be aliased, since $STV_3 = STV_1 = SRV_S$.*

Now suppose Algorithm 4.4 is applied, and the 3-color mapping given by Table 4.3 is obtained. If the same short S occurs, net n_3 will no longer be *aliased. In fact, from Table 4.3,* $SRV_3 = STV_3 = 001$*, which is different from* $SRV_S = STV_1 = 100$.

Finally, suppose Algorithm 4.4 is applied and a short $S' = \{e_{1,2}, e_{2,3}\}$ *with behavior* $S - D - 1$ *occurs. Then net* n_4 *will be aliased, since* $SRV_4 = STV_4 = 100$ *is the same vector as* $SRV_S = STV_1 = 100$.

The above example shows that with color mixing S-D-k aliasing is less likely to occur than with just graph coloring. On the other hand, the example also shows that the problem does not disappear: higher multiplicity faults can still cause S-D-k aliasing. In conclusion, color mixing can in fact improve diagnostic capability. Its possibilities will be further exploited in the next section, where a new structural diagnostic method will be presented.

Working independently, a research team from Texas A.M. University has achieved some interesting results on structural diagnosis (Feng et al., 1995; Chen and Lombardi, 1996; Liu et al., 1996; Salinas et al., 1996; Chen and Lombardi, 1998). In particular, they have proposed a better color mixing procedure (Chen and Lombardi, 1996; Chen and Lombardi, 1998), which is able to produce fewer colors than Algorithm 4.4 and can be summarized as follows. The first step is coloring graph G, just like in Algorithm 4.3. Then, the color mixing procedure is applied, which restricts its action to a subgraph T involving a net n_i, its neighbors denoted as *primary shorting nets* (PSNs), and the neighbors of its neighbors, denoted *secondary shorting nets* (SSNs). McBean and Moore have previously used the PSN and SSN concepts (McBean and Moore, 1993). Then, all possibilities for W-A/O structural confounding are enumerated in the subgraph T. If confounding problems are found in T, then extra colors are added to the initial color mapping; otherwise, no colors are added. That is, extra colors are added only when necessary, which leads to a global optimization in terms of the total number of colors used. In the end, the procedure produces fewer colors than Algorithm 4.4. Albeit solving a less important problem (structural confounding), this work makes a significant contribution: the need to indicate a subgraph dimension in which the diagnostic capability claims are valid. In the present work, the same conclusion has been reached independently and has been embodied in the concept of stain (de Sousa and Cheung, 1997b; de Sousa and Cheung, 1997a).

To conclude this section, the point made in Section 4.1.3, that color mixing methods can make up for the lost advantage from ignoring the visibility criterion in short extraction, is explained by the following example:

Example 4.9 *Suppose that the pin-to-net extraction mode is being used. Consider a pin belonging to a net* n_i *and, within the extraction radius* R*, two nets* n_j *and* n_k*. Now suppose that* n_j *is visible from* n_i*, and* n_j *is blocking the visibility of* n_k *from* n_i*. In this situation, the edges* $e_{i,j}$ *and* $e_{i,k}$ *will be extracted, but no edge* $e_{j,k}$ *will be extracted.*

When graph coloring is applied, it is possible that n_j *and* n_k *are assigned the same color:* $STV_j = STV_k$*. In this case, if there is a short* S *between* n_i *and* n_j*, that produces a response* $SRV_S = STV_j = STV_k$*, nets* n_j *and* n_k *become aliased. Had the visibility criterion been applied, it would have been realized that*

n_j is blocking the way between n_i and n_k. It would have also been realized that if a short $e_{i,k}$ were to be extracted, then a short $e_{j,k}$ should have been extracted too. That would have solved the problem.

The problem can also be solved by using color mixing techniques: color mixing will introduce an extra edge between n_j and n_k by applying the triangulation transformation. That avoids the n_j and n_k aliasing problem. Hence, the use of color mixing techniques can make up for the absence of the visibility criterion in the short extraction procedure.

4.4.4 Statistical color mixing vectors

The previous sections have shown that the structure of the graph used in preparing a structural diagnostic test set is the main factor in determining the diagnostic capability of the method. In that sense, Algorithm 4.4 improves upon Algorithm 4.3 by enhancing the structure of the adjacency graph G in order to create a graph G_2, which is then colored. Nonetheless, Algorithm 4.4 has two main limitations: (1) if faults have high multiplicities (for example, faults in an immature process) its diagnostic capability becomes inadequate; (2) the use of a walking sequence as a color coding scheme is too expensive in terms of STV length, and is unnecessary since confounding is unlikely. The statistical color mixing method introduced in this section will address both these limitations.

As concluded in Section 4.4.2, unless nets have unique STVs, no guarantee exists that a structural scheme will match the aliasing and confounding properties of the corresponding behavioral scheme. However, a theoretically weaker result can be established, which turns out to be very useful in practice. The following lemma (de Sousa and Cheung, 1997b; de Sousa and Cheung, 1997a) establishes this result:

Lemma 4.6 *Within a stain T, where all nets have unique STVs (colors), the diagnostic capability of a structural test set corresponds to the diagnostic capability of the behavioral test set that uses the same STV sequence.*

Proof: *According to the assumptions made so far, each net in a stain is a suspect net, and nets not in a stain are fault-free. Moreover, shorts between nets in different stains are impossible. Thence, because all nets in a stain T are being applied unique STVs, the problem corresponds to a behavioral diagnosis problem involving only the nets in the stain. Therefore, the diagnostic capability is the diagnostic capability of the behavioral test set.* ■

Lemma 4.6 shows that, only when stains involving nets of the same color do not occur, is it possible to make claims about the diagnostic capability of a structural test set. However, it cannot be guaranteed that such stains will not occur. If a large fault that involves many nets occurs, it is possible that a stain will contain nets of the same color. So, in the limit, the only way to guarantee that no nets of the same color will occur in a stain is to have as many colors as nets in a circuit. That is of course the same as behavioral diagnosis, which has already been studied in Chapter 3. An alternative method is to make the probability that a stain will contain nets assigned the same color as low as intended. This is the main idea behind the scheme described in this section — the statistical color mixing method.

Recall that a fault F is, in general, a set of single faults. Naturally, the occurrence probability $P(F)$ of F decreases with its multiplicity $r = \#F$ (detailed explanations will be given in Chapter 5). In practice, there is a value r^{max} such that, if $r > r^{max}$, then $P(F)$ can be neglected. If one can guarantee that no fault of multiplicity r^{max} can cause a stain containing nets of the same color, then it becomes possible to establish aliasing and confounding diagnostic properties. [8] One way to do that is to determine all multiplicity r^{max} faults, compute the stains they can produce, and make sure the nets in these stains are assigned unique colors. Unfortunately, even for a low r^{max}, the number of multiplicity r^{max} faults can be stupendous. Let M be the number of potential single shorts in a circuit. Obviously, M is equal to the number of edges in G, i.e., $M = \#E$. Then, the number of multiplicity r^{max} faults is

$$\sum_{r=0}^{r^{max}} \binom{N}{r} \binom{M}{r^{max} - r},$$

where $\binom{N}{r}$ is the number of multiplicity r opens and $\binom{M}{r^{max}-r}$ is the number of multiplicity $r^{max}-r$ shorts. Based on this observation, a technique to implicitly enumerate all these faults has been developed. This approach requires the following concepts:

Definition 4.15 (Edge-distance) *The edge-distance $ED(n_i, n_j)$, between two vertices n_i and n_j in a graph, is the number of edges in the shortest path between n_i and n_j.*

Definition 4.16 (Extent) *The extent $EXT(T)$ of a graph T is the maximum edge-distance in T.*

Example 4.10 *Consider the graph G in Figure 4.2(b). It can be easily seen that $ED(n_1, n_2) = 1$, $ED(n_1, n_3) = 2$, $ED(n_1, n_5) = 3$, etc. Also, $EXT(G) = \max_{n_i, n_j \in C} ED(n_i, n_j) = ED(n_1, n_5) = 3$.*

With the concepts of edge-distance and extent, the statistical color mixing method can now be explained. Let ε be called the *maximum stain extent.* The basic idea is to force any stain whose extent is equal or lower than ε to contain only nets assigned different colors. That is, within any such stain all STVs are unique. Then, coding the colors with some sequence denoted *coding_scheme*, its diagnostic properties can be guaranteed if the extent of any stain T never exceeds ε. The statistical color mixing method is outlined by Algorithm 4.5.

[8] It can be easily shown that, if no fault of multiplicity r_{max} can cause a stain in which there are nets assigned the same color, then the same is true for a fault of lower multiplicity $r < r_{max}$.

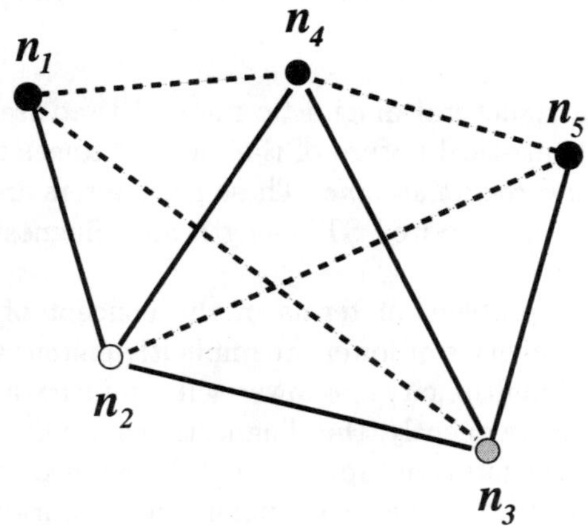

Figure 4.6. Graph G_2 and its 4-color mapping.

Algorithm 4.5 (generateStatColorMixVecs(G, ε, *coding_scheme*)) :

1. G_ε=*enhanceGraph(G,* ε*);* `/* See Appendix C */`
2. $CM(C)$ = *colorGraph(*G_ε*);* `/* See Appendix B */`
3. MTV = *codeColors(k, coding_scheme);* `/* codes the` k `colors using` *coding_scheme*`*/`
4. **return** MTV;

The complexity of Algorithm 4.5 is determined by the complexity of the graph enhancing step (line 1). The function *enhanceGraph()* takes as arguments a graph G and a maximum stain extent ε, and outputs an enhanced graph G_ε. To produce G_ε, the function *enhanceGraph()* adds extra edges between any two nets in G whose edge-distance is ε or lower. Algorithm C.1, described in Appendix C, has been used to implement this step. Its time complexity is $O(\varepsilon \times N^3)$. In Appendix C, a better algorithm for graph enhancing, Algorithm C.2, is also described. Algorithm C.2 has complexity $O(N \times D^\varepsilon)$.

Example 4.11 *Applying function Enhance() to graph G of Figure 4.2(b), with* $\varepsilon = 2$*, the 4-colored graph* G_2 *shown in Figure 4.6 is obtained.*

The diagnostic capability of a test set obtained with Algorithm 4.5 is established by the following theorem:

Theorem 4.6 *As long as* $EXT(T) \leq \varepsilon$ *for any stain* T*, Algorithm 4.5 produces schemes whose diagnostic capability corresponds to the diagnostic capability of the behavioral schemes obtained with the same STV sequence.*

Proof: *For a stain* T *such that* $EXT(T) \leq \varepsilon$*, it is true that* $\forall n_i, n_2 \in T : ED(n_i, n_j) \leq \varepsilon$*. Thus, any pair of nets* (n_i, n_j) *will be connected by an edge* $e_{i,j}$ *in* G_ε *after graph G is enhanced. That is, the nets in T will form a fully connected subgraph in* G_ε*. When* G_ε *is colored, each net in T will be assigned a* *unique STV from the STV sequence in use. The rest of the proof follows from Lemma 4.6.* ■

Algorithm 4.5 is a structural diagnostic method that uses statistical color mixing vectors. The statistical nature of the method comes from the selection of the parameters ε and *coding_scheme*. These parameters are chosen based on fault statistics, so that the desired STV length and diagnostic resolution can be met.

By formulating the problem in terms of the concept of stain extent, all possible faults of multiplicity ε or lower are implicitly enumerated. It can easily be shown that faults of multiplicity ε or lower will produce stains with an extent not greater than ε. Consequently, the diagnostic capability is guaranteed by Theorem 4.6. Even if the fault multiplicity slightly exceeeds ε, good diagnostic resolution can be provided. The level of diagnostic resolution will be assessed by means of a quantitative method, as will be explained in Chapter 5. If a new manufacturing process is being introduced, and low yields are expected, ε should be set to a conservative (high) value. As the process matures, and the production yield improves, ε may be decreased, which enables shortening the STV length.

Being able to choose an appropriate STV sequence for color coding is also important: since confounding is unlikely, a simply aliasing-free STV sequence (logarithmic complexity) usually ensures adequate diagnostic resolution. That is especially important in immature processes, where high fault multiplicities are expected, and ε must be set to a high value. Because the number of colors increases rapidly with ε (Section 4.4.3), using a logarithmic complexity color coding scheme is important for keeping the STVs short.

Algorithm 4.5 provides a solution that overcomes the limitations of Algorithm 4.4, presented in Section 4.4.3. The diagnostic ambiguity problem for high multiplicity faults is solved by means of setting ε to an adequate value. The STV length problem is solved by choosing a logarithmic complexity test sequence. Moreover, Algorith 4.4 is a special case of Algorithm 4.5 when $\varepsilon = 2$ and *coding_scheme* specifies the walking-1s sequence. In fact, the triangulation transformation (McBean and Moore, 1993) corresponds to constructing graph G_2. The more general Algorithm 4.5 provides mechanisms for optimizing the STV length and the diagnostic resolution, by allowing both ε and *coding_scheme* to be chosen by the user.

It should be noted that Algorithm 4.5 is a more general version of the *max-independence algorithm*(Yau and Jarwala, 1989). The max-independence algorithm uses a list of nets as its basic data structure, which it sorts according to adjacency criteria. The maximum stain extent ε is the equivalent of *maximum defect extent* E in the max-independence algorithm. Then, a diagonally independent set of STVs is applied to any consecutive $E + 1$ nets in the adjacency list. Algorithm 4.5 is more general than the max-independence algorithm, since a graph is better than a list at capturing neighboring relationships in a nearly planar layout. In a list, some neighboring relationships are not captured, which can invalidate the detection capability, as established by

Theorem 4.1. To avoid this problem, the max-independence algorithm does not dispense with STV uniqueness — each net receives a unique STV. In this way, detectability is guaranteed in the behavioral sense, as established by Theorem 3.1. Like Algorithm 4.5, the diagnostic resolution of the max-independence algorithm can be improved by setting E to a more conservative (high) value. Increasing E causes the STV length to increase rapidly, just like increasing ε in Algorithm 4.5. The max-independence algorithm could also be improved by replacing the prescribed diagonal independent test set with a cheaper logarithmic complexity scheme.

4.5 Summary

In this chapter, structural diagnosis of interconnect faults has been studied. In structural diagnosis, faults are diagnosed in terms of physically possible defects. The defects are previously extracted from the layout and their functional behavior is assumed to be given by a set of fault models.

An algorithm for PCB layout fault extraction has been presented. Open faults can be extracted in linear time, since they can only occur at solder joints. Shorts are more difficult to extract because it is necessary to search for the neighbor nets of each net. The time complexity of the fault extraction algorithm is $O(N^2)$. It has also been explained how the complexity of fault extraction could be optimized to $O(N \log N)$.

A method for structural diagnostic analysis has been presented. Although only physically possible defects are output as potential diagnoses, structural diagnosis is not necessarily less ambiguous than behavioral diagnosis. Diagnostic analysis should be preceded by a good diagnostic synthesis method, if low ambiguity is to be achieved. With a poor diagnostic synthesis method, structural diagnosis can be even worse than behavioral diagnosis. The time complexity of the structural diagnostic analysis algorithm is $O(N \log N)$.

Fault isolation in structural diagnosis is performed in a way similar to fault isolation in behavioral diagnosis. However, in structural fault isolation there is no need to look for the defects that correspond to the diagnosed faults, since those have been previously extracted and stored in a suitable data structure. Moreover, the same data structure can be used for implementing automatic repair techniques. Namely, the structural information is useful to guide repair robots to the defect locations. Thus, structural diagnosis can enable automatic repair technology.

Four distinct techniques for interconnect structural diagnostic synthesis have been studied: structural diagnosis with behavioral vectors, structural diagnosis with graph coloring vectors, structural diagnosis with color mixing vectors, and structural diagnosis with statistical color mixing vectors.

Structural diagnosis with behavioral vectors is an original technique proposed in this book. It uses structural analysis but maintains the behavioral synthesis of vectors. In this way, outstanding diagnostic resolution can be achieved because behavioral vectors become exceedingly powerful when combined with structural diagnostic analysis. In fact, structural information presents an opportunity for optimizing the STV length in diagnostic synthesis, which is totally missed by this approach.

The graph coloring method for diagnostic synthesis (Cheng et al., 1990) already uses structural information to optimize the STV length. The scheme enables a considerable reduction in STV length. However, it provides a very poor diagnostic capability, causing plenty of structural aliasing and confounding responses. Aliasing is caused by the S-D-k behavior, as first identified in this work. In behavioral diagnosis this problem does not occur, since S-D-k shorts can be diagnosed by vectors that target wired-logic shorts. However, in structural diagnosis S-D-k shorts elude vectors for W-A/O shorts, and create S-D-k aliasing. Additionally, S-D-k shorts can be quite common, and therefore cannot be ignored. As for the structural confounding problem, it had already been noticed by other authors (Lien and Breuer, 1991; McBean and Moore, 1993). However, structural confounding is less likely, since, for it to occur, it is necessary that two or more unconnected shorts cause a single stain in graph G. Unless behavioral vectors are used, neither the structural S-D-k aliasing problem nor the structural confounding problem can be solved completely.

The color mixing method (Lien and Breuer, 1991; McBean and Moore, 1993; Chen and Lombardi, 1998) uses more colors (more unique STVs) than the graph coloring method in order to reduce diagnostic ambiguity. The scheme reduces both the S-D-k aliasing problem and the structural confounding problem. However, the method has no flexibility for trading off diagnostic resolution and STV length.

The statistical color mixing method (de Sousa and Cheung, 1997b; de Sousa and Cheung, 1997a) provides a solution for balancing STV length and diagnostic resolution. The user can specify a subgraph maximum extent, within which any ambiguities are resolved, and the STV sequence to be used in color coding. The subgraph maximum extent is set according to statistics on fault multiplicity, which can be obtained from the manufacturing process. The STV sequence can be chosen freely, but usually a scheme of logarithmic complexity will be sufficient. The statistical color mixing method constitutes a practical and effective means to achieve high diagnostic resolution with affordable STV length.

5 DIAGNOSTIC RESOLUTION ASSESSMENT

In this chapter, a figure of merit for assessing the diagnostic resolution (DR) of diagnostic schemes is proposed. The method can be used to evaluate diagnostic schemes before selecting one for a particular electronic assembly (EA) of a given manufacturing process.

The traditional tool to evaluate interconnect diagnosis schemes is the aliasing/confounding framework. This framework provides a qualitative DR assessment, but cannot be reliably used to select a diagnostic scheme prior to its application. In contrast, the approach presented in this chapter quantitatively predicts DR for a statistically characterized process. The DR figure of merit is defined as a weighted average fault diagnosability. Each fault is weighted according to its occurrence probability and its individual diagnosability is computed by means of fault diagnosis simulation.

DR assessment by diagnosis simulation is a complex problem due to the combinatorial explosion caused by multiple faults. To restrict the complexity of the problem, faults above a certain multiplicity are excluded due to their low occurrence rate. From the remaining faults, a sample is extracted on which DR is evaluated. Some assumptions based on well known empirical facts allow treating short and open faults separately, which greatly simplifies the problem formulation.

This chapter is organized in four sections. In Section 5.1, background material is presented. In Section 5.2, a previously proposed stain-based DR assessment method is analyzed. In Section 5.3, an improved fault-based method is proposed. Section 5.4 presents a summary of the chapter.

5.1 Background

In this section, the background of the DR assessment problem is examined. Methods for actually measuring DR in a manufacturing environment are discussed. Some empirical facts concerning diagnostic capability are analyzed. The role of the aliasing/confounding framework in diagnostic capability assessment is put in perspective.

5.1.1 *Actual assessment*

In a production environment, various methods for monitoring the quality of a diagnostic methodology are possible. Here, a method based on the fault isolation efficiency FIE is described. According to Section 1.3, DR can be equated to FIE, because of the correspondence between diagnostic resolution and fault isolation time: if DR is high, the fault isolation time will be low, and vice-versa.

Suppose that some diagnostic scheme A can provide a perfect structural diagnosis F'' in the presence of a fault F. This means that $F'' = F$ (Section 3.1). With F'', a technician takes time τ_F to find the defects corresponding to fault F. Now suppose that another scheme B provides a complete and ambiguous diagnosis F'. If F' is a complete and ambiguous diagnosis, then $F \subset F'$ (Section 3.1). With F', the time taken to find the defects causing F is $\tau_{F'}$. Since diagnosis F'' is perfect and diagnosis F' is complete and ambiguous, then, on average, $\tau_F < \tau_{F'}$. Note that the execution times of procedures A and B are being neglected in τ_F and $\tau_{F'}$ compared to the time taken by the technician to visually inspect the EA and find fault F.

One possible definition of DR can be obtained by equating DR to the average fault isolation efficiency FIE:

$$DR = FIE = \frac{\langle \tau_F \rangle}{\langle \tau_{F'} \rangle}, \tag{5.1}$$

where $\langle \tau_F \rangle$ and $\langle \tau_{F'} \rangle$ represent the average fault isolation time with the perfect diagnosis F'' and with the diagnosis F', respectively. The above equation can be used to compute the cost of diagnosis and isolation (CDI) using Equation (1.4) of Section 1.3. An alternative definition of DR can be obtained by taking the average of the individual fault isolation efficiencies:

$$DR \doteq \langle \frac{\tau_F}{\tau_{F'}} \rangle. \tag{5.2}$$

The difference between the last two equations is subtle, but the values produced can be substantially different. The definition of DR used in this work is the one given by Equation (5.2). This definition has been the departure point for the DR figure of merit presented in Section 5.3. Later, it has been realized that the definition given by Equation (5.1) is probably more useful from a practical standpoint because it allows computing CDI. For historical reasons, the definition given by Equation (5.2) has been kept. However, besides differing in their values, the idea conveyed by the two equations is quite similar, and both can be used meaningfully. Thus, in either case, DR can be assessed using the following method:

1. For each faulty EA with a fault F, measure the time $\tau_{F'}$ to find the defects causing F with diagnosis F'.

2. After finding the fault, measure the time τ_F to find the defects with a perfect diagnosis $F'' = F$.

3. Having done the above for a sufficiently large number of faulty EAs, estimate DR using either Equation (5.1) or Equation (5.2).

5.1.2 *Empirical facts*

Efficient test and diagnosis methods are essential to guarantee the competitiveness of a manufacturing process. Thus, it is normal that details about the quality of the methods are kept confidential. Notwithstanding, several empirical facts are well known:

1. Schemes that simply guarantee fault detection are inadequate. In fact, with such low resolution methods, diagnosis becomes a tedious, lengthy and expensive process.

2. Aliasing-free schemes are essential to guarantee adequate diagnostic capability. Actually, they are the most popular schemes, especially the counting sequence plus complement scheme (Wagner, 1987).

3. Confounding-free schemes can often be dispensed with, and still an adequate diagnostic capability can be achieved. Either confounding does not occur frequently enough, or the long STV lengths discourage their use. [1]

4. Open faults often produce $\underline{0}/\underline{1}$ SRVs, are locally uncorrelated and can be unambiguously diagnosed. A S-A-0/1 aliasing-free sequence is enough to accurately diagnose open faults.

5. Short faults behave according to more sophisticated fault models, are often locally correlated and are more difficult to diagnose. Interconnect diagnosis schemes mainly focus on short faults.

These empirical facts have been observed with behavioral schemes that are commonly used in the industry. To our knowledge, empirical observations

[1] Recall that the STV length in confounding-free schemes is linear in the number of nets N, and that serial application with boundary-scan takes time $O(N^2)$, which may be excessive in many applications.

related to structural schemes have not been reported. The known empirical facts will be used to support the assumptions on which the DR assessment method described in Section 5.3 is based.

5.1.3 *Aliasing and confounding*

The interconnect diagnosis schemes described in two previous chapters have been analyzed with the aliasing/confounding framework. This framework enables *qualitative* evaluation of test sets and has originally been developed for comparing behavioral test sets among themselves. As shown in Chapter 4, the aliasing/confounding framework has also proved useful for comparing structural test sets among themselves. However, if we compare the aliasing/confounding evaluations given in Chapters 3 and 4 with the quantitative assessments given in Chapter 6, which are obtained with the method described in this chapter, it becomes clear that this framework is inadequate for comparing behavioral with structural schemes.

The evaluation enabled by the aliasing/confounding framework can be used to provide an indication about the quality of the scheme. However, it should not be the only criterion for selecting diagnostic schemes prior to their application. For some EA designs or manufacturing processes, a scheme as simple as the modified counting sequence may produce an acceptable DR. For other EAs or processes, it may be necessary to have an aliasing-free scheme such as the self-diagnosis scheme. Until the present work, there was no alternative to the aliasing/confounding framework other than performing an actual assessment on various schemes (Section 5.1.1) before choosing one.

Given the limitations of the aliasing/confounding framework, a quantitative DR figure of merit, whose value predicts the quality of a diagnostic scheme for a given design and process, has been a main goal in this research. The next two sections will present methods for quantitatively estimating DR based on defect statistics of the manufacturing process.

5.2 Stain-based assessment

In an earlier stage of this research, a DR assessment method based on the analysis of stains was proposed (Sousa et al., 1996b). The method is restricted to structural schemes and can only be used to evaluate short faults. The stain-based DR assessment method has been the first attempt to provide a quantitative DR measure. Here, the method will be quickly explained, in order to illustrate the difficulties that can arise when trying to set up a framework like this. More details about the method can be found elsewhere (de Sousa, 1996).

In the stain-based method a number of stains $T = (U, E')$ are injected in graph G and their diagnosabilities are computed. The concept of *stain diagnosability* d_T is given by

$$d_T = \frac{1}{m_T}, \tag{5.3}$$

where m_T is the *stain diagnosis ambiguity*, defined as the number of distinct sets of single shorts that can cause the stain. If $m_T = 1$ then $d_T = 1$, meaning that the stain can be diagnosed unambiguously; otherwise $m_T > 1$ and $d_T < 1$, which means that the diagnosis is ambiguous. Using stain diagnosabilities, the overall DR figure of merit in the stain-based method was defined (Sousa et al., 1996b) using the expression

$$DR = \sum_{i=1}^{N_s} d_{T_i} P_T(T_i), \tag{5.4}$$

where N_s is the total number of simulated stains, and $P_T(T_i)$ is the probability of occurrence of stain T_i given the circuit is faulty. $P_T(T_i)$ depends only on the stain size $\#U_i$, and decreases with it. The maximum size of the stains to be simulated is determined so that the exclusion of stains of greater sizes does not cause a significant error in DR. The probability function $P_T(T_i)$ is normalized so that $\sum_{i=1}^{N_s} P_T(T_i) = 1$, which, together with $0 < d_T \leq 1$, implies that $0 < DR \leq 1$. If all stain diagnosabilities are evaluated to 1 then $DR = 1$; otherwise, DR is lower than 1.

The stain-based method for DR assessment ranks structural schemes according to their diagnostic capability. This ranking agrees with the more recent fault-based approach that will be presented in Section 5.3. However, at the time when the stain-based approach was devised, the concern of it being a measure of fault isolation efficiency was not there. Despite being able to compare two schemes in relative terms, the absolute DR values obtained with the two methods may differ significantly.

Another achievement of the stain-based approach is the insight it has provided on confounding. After injecting a stain in graph G, all shorts that can cause the stain are determined, including all possible confounding shorts. In this way, confounding has been studied for several diagnostic schemes. Between confounding and confounding-free schemes, only small DR differences have been observed, which leads to the conclusion that confounding is perhaps tolerable, in agreement with empirical fact 3 of Section 5.1.2.

Despite providing a new way of assessing DR, several problems have been encountered in the stain-based approach: (1) the stain size probability distribution depends on the diagnostic scheme under evaluation and not exclusively on the stain size; (2) open faults are ignored, which distorts DR; (3) the method cannot be used to compare structural with behavioral schemes.

The dependence of the stain size on the diagnostic scheme is the following: given the same faulty EA, the stains produced by a scheme of higher diagnostic capability are likely to be smaller than the stains produced by a lower capability scheme. This means that when two schemes are compared with the same stain, the lower diagnostic capability method is always favored. Nevertheless, the higher capability scheme will still be less ambiguous, which gives the stain-based method the ability to compare schemes in relative terms. The results obtained by de Sousa and Cheung with the stain-based approach have this problem (Sousa et al., 1996b).

By considering only short faults and ignoring open faults, the stain-based method makes the overall DR somewhat conservative. According to empirical facts 4 and 5 of Section 5.1.2, open faults are in general easier to diagnose than short faults, and therefore ignoring them causes DR to be underestimated.

Since it does not model behavioral diagnostic ambiguity, the stain-based method cannot be used to compare structural with behavioral schemes. In structural diagnosis, the diagnostic ambiguity is modeled by m_T, obtained by analysis of the stain T. However, behavioral diagnoses may have no representation in the adjacency graph, which makes it difficult to model diagnostic ambiguity.

5.3 Fault-based assessment

In this section, a different definition of diagnostic resolution (DR) and a DR assessment method are given (de Sousa and Cheung, 1997a). The definition of DR tries to reflect the time for locating actual faults, regardless of the type of approach: behavioral or structural. The assessment method is based on statistical fault diagnosis simulation.

In the fault-based approach, faults are injected in G and the stains produced are analyzed. This is exactly the inverse procedure of the stain-based method, where stains are injected and the faults that can cause them are determined. An advantage of the fault-based method is that faults can be generated according to statistics obtained from the manufacturing process, which, unlike stain statistics, are independent of the diagnostic scheme being assessed.

The derivation of the fault-based approach to DR assessment comprises the following steps: the definition of a probabilistic DR measure, the development of fault probability models and fault sampling techniques, the calculus of individual fault diagnosabilities and an algorithm for fault diagnosis simulation.

5.3.1 *Probabilistic resolution*

In this section, a probabilistic formulation of DR is presented. The formulation is based on computing the ***individual fault diagnosabilities*** $d_{i,j}^{r_s,r_o}$ of each fault $F_{i,j}^{r_s,r_o}$, where the superscript r_s, r_o indicates that the fault has r_s single shorts and r_o single opens and the subscript i, j indicates that among all possible $\binom{M}{r_s}$ combinations of multiplicity r_s shorts, and among all possible $\binom{N}{r_o}$ combinations of multiplicity r_o single opens, fault $F_{i,j}^{r_s,r_o}$ has the combination of shorts i and the combination of opens j. The DR figure of merit is given by

$$DR = \sum_{r_s=0}^{M} \sum_{r_o=0}^{N} \sum_{i=1}^{\binom{M}{r_s}} \sum_{j=1}^{\binom{N}{r_o}} d_{i,j}^{r_s,r_o} P_F(F_{i,j}^{r_s,r_o} | r_s > 0 \vee r_o > 0), \tag{5.5}$$

where $P_F(F_{i,j}^{r_s,r_o} | r_s > 0 \vee r_o > 0)$ is the occurrence probability of fault $F_{i,j}^{r_s,r_o}$ under the condition $r_s > 0 \vee r_o > 0$. The condition $r_s > 0 \vee r_o > 0$ indicates that some fault has occurred. Thus, the conditional probability $P_F(F_{i,j}^{r_s,r_o} | r_s >$

$0 \vee r_o > 0)$ can be given by

$$P_F(F_{i,j}^{r_s,r_o}|r_s > 0 \vee r_o > 0) = \begin{cases} 0, & r_s = r_o = 0, \\ \frac{P_F(F_{i,j}^{r_s,r_o})}{1-Y}, & \text{otherwise,} \end{cases} \tag{5.6}$$

where $P_F(F_{i,j}^{r_s,r_o})$ is the occurrence probability of fault $F_{i,j}^{r_s,r_o}$ and Y is the yield, i.e., the probability that no fault has occurred. Since $P_F(F_{i,j}^{r_s,r_o}|r_s > 0 \vee r_o > 0)$ is a probability function,

$$\sum_{r_s=0}^{M} \sum_{r_o=0}^{N} \sum_{i=1}^{\binom{M}{r_s}} \sum_{j=1}^{\binom{N}{r_o}} P_F(F_{i,j}^{r_s,r_o}|r_s > 0 \vee r_o > 0) = 1. \tag{5.7}$$

The calculus of the individual fault diagnosabilities will be carried out in Section 5.3.5. As will be shown there, the individual fault diagnosabilities $d_{i,j}^{r_s,r_o}$ verify the condition $0 < d_{i,j}^{r_s,r_o} \leq 1$. Thus, according to Equation (5.5), DR verifies the condition $0 < DR \leq 1$; DR evaluates to 1 when all fault diagnosabilities $d_{i,j}^{r_s,r_o}$ evaluate to 1 and is lower than one if some fault diagnosabilities $d_{i,j}^{r_s,r_o}$ are lower than 1. Computing DR by Equation (5.5) is not feasible because it involves determining the individual fault diagnosabilities $d_{i,j}^{r_s,r_o}$ of an exceedingly high number of faults. To overcome this difficulty, the diagnosabilities $d_{i,j}^{r_s,r_o}$ are computed only for a fault sample. Fault sampling will be discussed in Section 5.3.3. Using the Bayes rule, the fault occurrence probability $P_F(F_{i,j}^{r_s,r_o})$ is expressed by

$$P_F(F_{i,j}^{r_s,r_o}) = P_F(F_{i,j}^{r_s,r_o}|r_s, r_o) P_{s,o}(r_s, r_o), \tag{5.8}$$

where $P_{s,o}(r_s, r_o)$ is the probability that r_s single shorts and r_o single opens will occur. Now, the following assumption is introduced:

Assumption 5.1 *Short faults occur independently of open faults, that is, r_s and r_o are independent random variables.*

This assumption is reasonable because investigations have shown no evidence of shorts and opens being statistically correlated. According to Assumption 5.1, Equation (5.8) can be written in the form

$$P_F(F_{i,j}^{r_s,r_o}) = P_{Fs}(F_i^{r_s}|r_s) P_s(r_s) \cdot P_{Fo}(F_j^{r_o}|r_o) P_o(r_o), \tag{5.9}$$

where $P_{Fs}(F_i^{r_s}|r_s)$ is the occurrence probability of short $F_i^{r_s}$ given that r_s shorts have occurred, $P_s(r_s)$ is the occurrence probability of r_s shorts, $P_{Fo}(F_j^{r_o}|r_o)$ is the occurrence probability of short $F_i^{r_o}$ given that r_o opens have occurred, and $P_o(r_o)$ is the occurrence probability of r_o opens. Substituting the above equation in Equation (5.6), and then substituting the resulting expression in Equation (5.5), the following equation is obtained:

$$\begin{aligned} DR = \tfrac{1}{1-Y} \Big[& \textstyle\sum_{r_s=0}^{M} P_s(r_s) \sum_{i=1}^{\binom{M}{r_s}} P_{Fs}(F_i^{r_s}|r_s) \\ & \times \textstyle\sum_{r_o=0}^{N} P_o(r_o) \sum_{j=1}^{\binom{N}{r_o}} d_{i,j}^{r_s,r_o} P_{Fo}(F_j^{r_o}|r_o) - Y \Big]. \end{aligned} \tag{5.10}$$

The advantage of expressing DR by the above equation is that it allows the use of separate distributions for short and open faults. As will soon become evident, formulating DR in this way greatly facilitates the development of the DR assessment method.

5.3.2 Fault probabilities

In PCB processes, the *negative binomial distribution* with parameters λ and α has been found to be an appropriate model for the fault occurrence probability (Tegethoff, 1994a; Tegethoff, 1994b). The probability that r faults will occur, $P(r)$, can also be called the *fault multiplicity distribution*. The negative binomial formulation has also been used in IC yield modeling (Stapper et al., 1983; Seth and Agrawal, 1984; de Sousa and Agrawal, 2000), and, to describe a spatial distribution of retail shops in a city (Rogers, 1974). Its expression is

$$P(r) = \frac{\Gamma(\alpha + r)}{r!\,\Gamma(\alpha)} \frac{(\frac{\lambda}{\alpha})^r}{(1 + \frac{\lambda}{\alpha})^{r+\alpha}}, \tag{5.11}$$

where, λ is the average number of occurrences of a certain phenomenon in a predefined space interval, and α is the *clustering parameter*, which measures the tendency for spatial clustering of the occurrences. In the context of PCB solder faults, λ is the average number of faulty joints per EA, and α is the fault clustering parameter. [2] The process yield Y for a negative binomial fault multiplicity distribution is given by

$$Y = P(0) = (1 + \frac{\lambda}{\alpha})^{-\alpha}. \tag{5.12}$$

Both λ and α are positive constants. A low value of α corresponds to a strong clustering effect; as α increases the clustering effect weakens. In the limit ($\alpha \to \infty$), Equation (5.11) tends to a Poisson distribution with parameter λ:

$$P(r) = \frac{\lambda^r e^{-\lambda}}{r!}. \tag{5.13}$$

For a Poisson fault multiplicity distribution the yield Y reduces to

$$Y = P(0) = e^{-\lambda}. \tag{5.14}$$

Both Equation (5.12) and Equation (5.14) are well known formulae from the the IC yield theory (Stapper et al., 1983). The distribution $P(r)$ is defined for any integer $r \geq 0$. In strict theoretical terms, $P(r)$ should be defined only for $0 \leq r \leq N_F$, where N_F is the total number of possible faults. However, since N_F is usually large, $P(r)$ can be defined for $0 \leq r \leq +\infty$, without significant loss of accuracy. It is important is that $P(r)$ is accurate for the low values of r; since $P(r)$ decreases rapidly with increasing r, it becomes practically zero for values of r that are much lower than N_F, and can be neglected.

[2] Weak clustering means that faults will distribute themselves among as many boards as possible, whereas strong clustering means that many faults will occur only on a few boards.

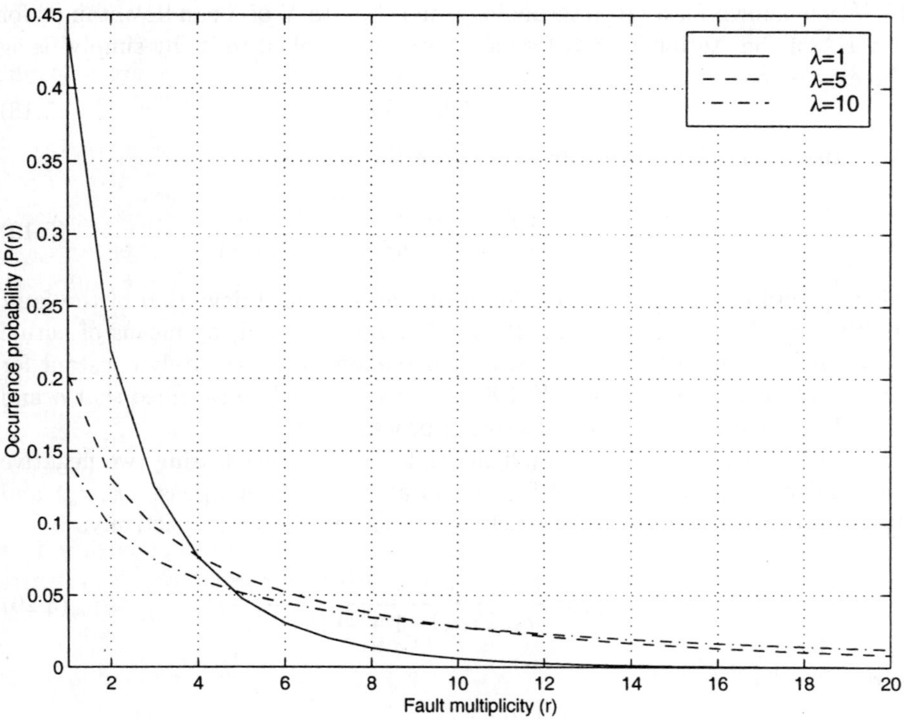

Figure 5.1. Fault probability $P(r)$ versus fault multiplicity r in a faulty EA.

In diagnosis, it is more convenient to work with the fault multiplicity distribution in faulty EAs, since diagnosis is only performed on faulty EAs. That distribution is given by

$$P(r|r > 0) = P(r)/(1 - Y). \tag{5.15}$$

This distribution is shown in Figure 5.1 for three values of λ and a fixed value of α. It can be observed that when λ increases the whole distribution levels down, making the probability more even across the fault multiplicity axis. A similar effect can be observed if α is decreased while keeping λ constant, which suggests a certain interchangeability in the roles of the two parameters. Nevertheless, the two parameters are necessary for modeling clustered distributions.

According to the physical meaning of λ, an estimate λ' of λ can be obtained from the manufacturing process using the formula

$$\lambda' = \frac{\#\text{faults found}}{\#\text{EAs tested}}. \tag{5.16}$$

Parameter λ characterizes the EA and not the technology. The parameter that characterizes the technology is the *solder fault rate* SFR, given by

$$SFR = \frac{\lambda}{N_F}. \tag{5.17}$$

If SFR is known for a given technology, an estimate λ' of λ can be obtained for any EA of that technology, before any tests are applied to it, by simply using the expression

$$\lambda' = SFR \times N_F. \tag{5.18}$$

An estimate α' of α can be obtained using the formula

$$\alpha' = \frac{\mu^2}{\sigma^2 - \mu^2}, \tag{5.19}$$

where μ and σ are estimates of the mean and standard deviation of the fault multiplicity data (Cunningham, 1990). Tegethoff showed, by means of actual measurements in an SMT process, that α remains approximately constant for a range of board sizes (Tegethoff, 1994a). Hence, it will be assumed that α and SFR fully characterize a manufacturing process.

According to Section 5.3.1, fault multiplicity is modeled using two negative binomial distributions: $P_s(r_s)$ for shorts and $P_o(r_o)$ for opens. $P_s(r_s)$ and $P_o(r_o)$ are characterized by parameters λ_s, α_s and λ_o, α_o, respectively:

$$P_s(r_s) = \frac{\Gamma(\alpha_s + r_s)}{r_s!\,\Gamma(\alpha_s)} \frac{(\frac{\lambda_s}{\alpha_s})^r_s}{(1 + \frac{\lambda_s}{\alpha_s})^{r + \alpha_s}}, \tag{5.20}$$

and

$$P_o(r_o) = \frac{\Gamma(\alpha_o + r_o)}{r_o!\,\Gamma(\alpha_o)} \frac{(\frac{\lambda_o}{\alpha_o})^r_o}{(1 + \frac{\lambda_o}{\alpha_o})^{r + \alpha_o}}. \tag{5.21}$$

The probability that there are no short faults will be called the *shorts yield* Y_s which, according to Equation (5.12), is given by

$$Y_s = \left(1 + \frac{\lambda_s}{\alpha_s}\right)^{-\alpha_s}. \tag{5.22}$$

The probability that there are no open faults will be called the *opens yield* Y_o, given by

$$Y_o = \left(1 + \frac{\lambda_o}{\alpha_o}\right)^{-\alpha_o}. \tag{5.23}$$

The total yield of the process can be obtained as

$$Y = Y_s Y_o = \left(1 + \frac{\lambda_s}{\alpha_s}\right)^{-\alpha_s} \left(1 + \frac{\lambda_o}{\alpha_o}\right)^{-\alpha_o}. \tag{5.24}$$

Similar to Equation (5.18), an estimate λ_s' of λ_s can be obtained as

$$\lambda_s' = SSR \times N_F \tag{5.25}$$

where SSR is the *solder short rate*. In the same way an estimate λ_o' of λ_o is given by

$$\lambda_o = SOR \times N_F, \tag{5.26}$$

where SOR is the *solder open rate.* The parameters SSR and SOR can be used to characterize the process, instead of the parameter SFR given before. Note that,

$$SFR = SSR + SOR. \tag{5.27}$$

Multiplying both sides of the above equation by N_F, the equation that relates λ, λ_s and λ_o is obtained:

$$\lambda = \lambda_s + \lambda_o. \tag{5.28}$$

Parameters α_s and α_o can be estimated from distributions $P_s(r_s)$ and $P_o(r_o)$ using the estimator given by Equation 5.19. Hence, when DR is formulated in terms of the distributions $P_s(r_s)$ and $P_o(r_o)$, the parameters that characterize the process are SSR, SOR, α_s and α_o.

5.3.3 Fault sampling

DR as given by Equation (5.10) is to be estimated using a fault sample, since it is impractical to simulate the whole population of possible faults. In this section, the problem of how to perform fault sampling is addressed. First it is shown that, although their contribution to DR must be accounted for, open faults need not be sampled. Second, it is explained how the multiplicity and topology of shorts are restricted, in order to simplify fault sampling and improve the accuracy of DR.

Open faults. Empirical fact 4 of Section 5.1.2 indicates that open faults are often locally uncorrelated and easy to diagnose. Opens can cluster at the global EA level due to process variations, in the sense that they may concentrate in a few EAs of a production lot. However, they do not tend to form local clusters in the EAs themselves. If multiple opens occur, they are most likely scattered in the layout and usually exhibit S-A-0/1 behavior. These facts lead to the following assumption:

Assumption 5.2 *Open faults do not cause diagnostic ambiguity.*

In fact, isolated opens having S-A-0/1 behavior can be unambiguously diagnosed by most test sets, behavioral or structural. Note that most diagnostic schemes are S-A-0/1 aliasing-free, which is enough to provide adequate diagnosis of these faults (Sections 3.1 and 4.2).

A fault $F_{i,j}^{r_s,r_o}$ can be decomposed in two component faults: a short fault $F_i^{r_s}$ with diagnosability $d_i^{r_s}$ and an open fault $F_j^{r_o}$ with diagnosability $d_j^{r_o}$. Since diagnostic ambiguity is being modeled in terms of the fault isolation time, Assumption 5.2 implies that a short fault isolation time can be expected for open faults, which perhaps can be neglected when compared to the fault isolation time of short faults. Since diagnosability is the inverse of diagnostic ambiguity, this hypothesis corresponds to the following simplification

$$d_{i,j}^{r_s,r_o} = \begin{cases} 1, & r_s = 0, \\ d_i^{r_s}, & r_s > 0. \end{cases} \tag{5.29}$$

This equation states that if a fault $F_{i,j}^{r_s,r_o}$ has a component short fault $F_i^{r_s}$, i.e., if $r_s > 0$, then its diagnosability $d_{i,j}^{r_s,r_o}$ is solely the diagnosability $d_i^{r_s}$ of $F_i^{r_s}$. If it is composed solely of open faults ($r_s = 0$), then its diagnosability is 1. With this simplification, Equation (5.10) drastically reduces to

$$DR = \frac{1}{1-Y}\left(\sum_{r_s=0}^{M} P_s(r_s) \sum_{i=1}^{\binom{M}{r_s}} d_i^{r_s} P_{Fs}(F_i^{r_s}|r_s) - Y\right). \tag{5.30}$$

According to the above equation, open faults do not need to be considered when evaluating DR. Nonetheless, their contribution is still encapsulated in Equation (5.30). To show this, Equation (5.30) is rewritten in the form

$$DR = \frac{1}{1-Y}\left(\sum_{r_s=0}^{M} d^{r_s} P_s(r_s) - Y\right), \tag{5.31}$$

where d^{r_s} is the *average diagnosability of multiplicity* r_s *shorts*, given by

$$d^{r_s} = \sum_{i=1}^{\binom{M}{r_s}} d_i^{r_s} P_{Fs}(F_i^{r_s}|r). \tag{5.32}$$

It can be shown that

$$\sum_{r_s=0}^{M} d^{r_s} P_s(r_s) = Y_s + (1 - Y_s) DR_s, \tag{5.33}$$

where DR_s is the *shorts diagnostic resolution*, given by

$$DR_s = \sum_{r_s=1}^{M} d^{r_s} P_s(r_s|r_s > 0). \tag{5.34}$$

Substitution of Equation (5.33) in Equation (5.31) produces the expression

$$DR = \gamma_o + (1 - \gamma_o) DR_s, \tag{5.35}$$

where

$$\gamma_o = \frac{Y_s - Y}{1 - Y}. \tag{5.36}$$

Equation (5.35) shows that there are two distinct contributions to DR: the fault fraction γ_o contributes with diagnosability 1 to DR and the fault fraction $1 - \gamma_o$ contributes an average diagnosability DR_s. What is the meaning of γ_o? It can be shown that

$$\gamma_o = \sum_{r_o=1}^{N} P_{s,o}(0, r_o|r_s > 0 \vee r_o > 0), \tag{5.37}$$

which signifies that γ_o is the probability that there are only open faults, given the EA is faulty. The γ_o fault fraction can be quite significant since $P_s(r_s)$ and $P_o(r_o)$ are independent negative binomial distributions. With these distributions, the probability that there are only opens or only shorts is quite high when compared to the probability of mixed open-short faults. If $DR_s = 1$, it means that shorts (as well as opens) can be unambiguously diagnosed, which causes $DR = 1$. However, if $DR_s = 0$, only open faults contribute to DR, making $DR = \gamma_o$. If no shorts occur ($Y_s = 1$), then $\gamma = 1$ and $DR = 1$. On the other hand, if no open faults occur ($Y_o = 1$), then $Y = Y_s$, $\gamma_o = 0$ and $DR = DR_s$.

Short faults. Since $P_s(r_s)$ is a negative binomial formula, it typically decreases rapidly with increasing r_s, suggesting that shorts above a certain multiplicity r_s^{max} can be neglected. The *maximum short multiplicity* r_s^{max} should be set high enough so that the error ϵ caused in DR does not exceed a certain pre-imposed value. Imposing a maximum error ϵ to DR corresponds to imposing a maximum error ϵ_s to DR_s. According to Equation (5.35), ϵ_s and ϵ are related by the following equation:

$$\epsilon_s = \frac{\epsilon}{1-\gamma_o}. \tag{5.38}$$

When the summation to M in Equation (5.34) is replaced with a summation to r_s^{max}, the resulting error must not exceed ϵ_s. Then, the following inequality must hold:

$$\sum_{r_s=r_s^{max}+1}^{M} d^{r_s} P_s(r_s|r_s>0) < \epsilon_s. \tag{5.39}$$

A conservative value of r_s^{max} that satisfies the above inequality can be obtained by setting the diagnosabilities d^{r_s} to 1 and solving the inequality for the minimum r_s^{max}:

$$r_s^{max} = \min m : \sum_{r_s=1}^{m} P_s(r_s|r_s>0) > 1-\epsilon_s. \tag{5.40}$$

To compute r_s^{max} using the above equation one just needs to know the parameters λ_s and α_s of distribution $P_s(r_s)$. According to Section 5.3.2, r_s^{max} is expected to increase if λ_s increases or α_s decreases. This shows, as it is intuitive, that the maximum short multiplicity r_s^{max} increases with both the circuit size and the clustering effect.

An estimate of the average diagnosability d^{r_s} of multiplicity r_s shorts is computed using the multiplicity r_s shorts in the fault sample. A criterion for short sampling is derived from empirical fact 5 of Section 5.1.2, which states that multiple shorts are often locally correlated. Locally correlated shorts occur, for example, when a large solder blob causes several nets in a PCB to be shorted. This observation leads to the following assumption.

Assumption 5.3 *The diagnosability of short faults is mainly determined by the diagnosability of the shorts that form connected subgraphs in G.*

According to this assumption, shorts that form connected subgraphs in G provide a better estimate of the diagnosabilities d^{r_s} than shorts sampled at random. Thus, the short sampling criterion states that only shorts that form connected subgraphs should be considered. There is however a problem with this criterion: since only unconnected short subgraphs can confound, confounding shorts will not be sampled. In structural diagnosis this may not be a problem since the probability of confounding is low. Note that only if the short subgraphs are adjacent (Definition 4.14), and have the same response, will confounding occur. In contrast, in behavioral diagnosis any shorts that have the same response will confound. Hence, it is possible that evaluating DR exclusively with shorts that form connected subgraphs leads to a slightly overestimated DR value.

The topology of the short subgraphs may also influence DR. Many short subgraphs will have a linear shape, especially if short extraction is performed in the pin-to-pin mode (Section 4.1.2). In this mode, most shorts involve consecutive pins of the same IC. Sometimes though, other subgraph shapes may occur. For example, a triangular subgraph shape will occur in the following situation: three nets n_i, n_j and n_k are connected to three successive pins of an IC, so that the single shorts $e_{i,j}$ and $e_{j,k}$ are possible; n_i and n_k are connected to two successive pins of another IC, so that short $e_{i,k}$ is possible. Thence, a triangular short subgraph $S = (\{n_i,\ n_j,\ n_k\},\ \{e_{i,j},\ e_{j,k},\ e_{i,k}\})$ is possible. When there is a choice among subgraph shapes, the shape that maximizes the number of nets in the short will be preferred. It should be said that this is a conservative choice, which may cause DR to be slightly underestimated. [3]

5.3.4 Overall DR estimate

It has been shown how the short sampling assumptions have led to constraining the multiplicity and topology of the sampled shorts. Those assumptions have been carefully justified since fault sampling is a very sensitive aspect of the fault-based approach to DR assessment. The fault sample obtained using the techniques described will now be used to estimate an overall DR value. First, estimates $d^{r_s\prime}$ of the diagnosabilities d^{r_s} that appear in Equation (5.34) are computed by the expression

$$d^{r_s\prime} = \frac{1}{N_{r_s}} \sum_{i=1}^{N_{r_s}} d_i^{r_s}, \tag{5.41}$$

[3] In behavioral diagnosis, more nets in a short means more diagnostic ambiguity, as will be shown in Section 5.3.5. In structural diagnosis, the more nets are involved in a short the more likely the short subgraph will contain nets assigned the same color. Recall from Chapter 4 that nets assigned the same color are assigned the same STV, and that diagnosis becomes more ambiguous if shorts contain nets assigned the same STV.

where N_{r_s} is the number of sampled faults of multiplicity r_s. As will be explained in Section 5.3.6, $d^{r_s}{}'$ is determined by diagnosis simulation. It has been empirically verified that $N_{r_s} = 1000$ is usually enough to produce an accurate $d^{r_s}{}'$ estimate, independent of the circuit size. Then, an estimate DR'_s of the shorts diagnostic resolution DR_s can be obtained by substituting d^{r_s} with $d^{r_s}{}'$ in Equation (5.34):

$$DR'_s = \sum_{r_s=1}^{r_s^{max}} d^{r_s}{}' P_s(r_s | r_s > 0). \tag{5.42}$$

An estimate DR' of the overall diagnostic resolution DR is obtained by substituting DR_s with DR'_s in Equation (5.35):

$$DR' = \gamma_o + (1 - \gamma_o) DR'_s. \tag{5.43}$$

5.3.5 *Individual diagnosabilities*

Now, the calculus of the individual short diagnosabilities $d_i^{r_s}$ is addressed. The idea is that the individual short diagnosability $d_i^{r_s}$ of a short $F_i^{r_s}$ should reflect the individual fault isolation efficiency $\tau/\tau_{F'}$ for diagnosis F'. It is assumed that the fault isolation time $\tau_{F'}$ is proportional to the number of single shorts in F'. Thus, a possible expression for $d_i^{r_s}$ is

$$d_i^{r_s} = \langle \frac{r_s}{r'_s} \rangle, \tag{5.44}$$

where r'_s is the number of single shorts in diagnosis F', and $\langle r_s/r'_s \rangle$ is the average of r_s/r'_s for all the N_b possible behaviors of fault $F_i^{r_s}$. If each behavior b is characterized by a probability $P_b(b)$, then

$$d_i^{r_s} = \sum_{b=1}^{N_b} [d_i^{r_s}]_b P_b(b), \tag{5.45}$$

where $[d_i^{r_s}]_b$ is the diagnosability of short $F_i^{r_s}$ for behavior b. The distribution $P_b(b)$ can be statistically obtained from the manufacturing process. The diagnosability $[d_i^{r_s}]_b$ for behavior b is given by

$$[d_i^{r_s}]_b = \frac{r_s}{[r'_s]_b}, \tag{5.46}$$

where $[r'_s]_b$ is the number of diagnosed shorts for behavior b. In behavioral diagnostic analysis, it is assumed that shorts between any two nets in a set of suspect nets U are possible. Thus,

$$[r'_s]_b = \binom{\#U}{2} = \frac{(\#U)^2 - \#U}{2}. \tag{5.47}$$

In structural diagnostic analysis, it is assumed that only the shorts in a stain $T = (U, E')$ are possible. Since the number of possible single shorts in T is the number of edges of T, therefore

$$[r'_s]_b = \#E'. \tag{5.48}$$

According to the type of diagnostic analysis, behavioral or structural, Equation (5.47) or Equation (5.48), respectively, should be substituted in Equation (5.46) to yield the value of $[d_i^{r_s}]_b$.

5.3.6 *Diagnosis simulation*

Based on the theory presented in the previous sections, an algorithm for diagnosis simulation has been developed. The algorithm takes as input a test set *MTV*, the diagnosis mode (behavioral or structural), the adjacency graph G and the following statistical parameters: λ_s, α_s, λ_o, α_o, ϵ and N_{r_s}. Then, the *DR* figure of merit is computed using the following procedure:

Algorithm 5.1 (diagSim(MTV, $MODE$, G, λ_s, λ_o, α_s, α_o, ϵ, N_{r_s}**))** :
1. r_s^{max} = *computeMaxShortMult*(λ_s, α_s, λ_o, α_o, ϵ); `/*Equation (5.40)*/`
2. **for** $r_s = 1$ **to** r_s^{max} **do** {
3.   **for** $i = 1$ **to** N_{r_s} **do** {
4. $F_i^{r_s}$ = *injectShort*(G, r_s);
5.   **for** `/*behavior*/` $b = 1$ **to** $b = N_b$ **do** {
6.   **if**(*MODE* == *BEHAVIORAL*) {
7. U = *computeSetOfSuspectNets*($F_i^{r_s}$, b, MTV);
8. r'_s = *computeDiagnosedShorts*(U); `/*Equation (5.47)*/` }
9.   **if**(*MODE* == *STRUCTURAL*) {
10. T = *computeStain*(G, $F_i^{r_s}$, b, MTV);
11. $[r'_s]_b$=*computeDiagnosedShorts*(T); `/*Equation (5.48)*/` }
12. $[d_i^{r_s}]_b$=*computeDiagnosability*(r_s, $[r'_s]_b$); `/*Equation (5.46)*/` }
13. $d_i^{r_s}$ = *computeIndividualDiagnosability*(); `/*Equation (5.45)*/` }
14. $d^{r_s\prime}$ = *computeAverageDiagnosability*(); `/*Equation (5.41)*/` }
15. DR'_s = *computeShortsDR*(λ_s, α_s); `/*Equation (5.42)*/`
16. DR' = *computeDR*(λ_s, α_s, λ_o, α_o, DR_s); `/*Equation (5.43)*/`
17. **return** *DR'*;

Algorithm 5.1 has complexity $O(r_s^{max} \times N_{r_s} \times N_b \times X)$, where X is a term that depends on the diagnosis mode. In the behavioral mode, $X = p \times N$ due to the computation of the set of suspect nets U in line 7. [4] In the structural mode, $X = p \times r_s^{max} \times D$ as a result of the computation of stain T in line 10. [5] In behavioral diagnosis, X depends on the EA size: it depends on N, which

[4] Algorithm 3.1 in Section 3.1 explains how to compute the set of suspect nets U.
[5] The stain computation procedure is described in Section 4.2 by Algorithm 4.2.

is directly related to the EA size, and depends on p, itself a function of N. [6] In structural diagnosis, X does not depend on the EA size: D depends on the structure of graph G, p depends on the diagnostic method and on the graph structure and r_s^{max} depends solely on fault statistics. [7] Therefore, the complexity of Algorithm 5.1 is $O(r_s^{max} \times N_{r_s} \times N_b \times p \times N)$ in the behavioral mode and $O((r_s^{max})^2 \times N_{r_s} \times N_b \times p \times D)$ in the structural mode.

The efficiency of diagnosis simulation can be significantly improved if the specifics of each test set are exploited. In behavioral diagnosis, if the test set is aliasing free, U does not need to be computed because it corresponds to the set of shorted nets V. In this case, simulation of the short behaviors is unnecessary since $[r'_s]_b$ is independent of the fault behavior b and always given by $r'_s = \binom{\#V}{2}$. Hence, the complexity of the algorithm can be reduced to $O(r_s^{max} \times N_{r_s})$. If the test set is an aliasing scheme, i.e., the counting sequence or the modified counting sequence, an aliased net n_j such that $i = SRV_S{+}1$ (counting sequence) or $i = SRV_S$ (modified counting sequence) may occur. [8] The computational cost of converting SRV_S to an integer, so that it can be compared with $i-1$ or i, is $O(p)$. For n_i to be aliased, two conditions must be verified: (1) $i < N$ for the counting sequence ($i \leq N$ for the modified counting sequence); (2) $n_i \notin V$. Condition (1) simply checks if SRV_S matches one of the STVs in the test set. Condition (2) verifies if n_i is not in the original set of shorted nets V (if it were, then no aliasing would occur, since all diagnosed nets would be faulty). To verify (2) at most r_s^{max} comparisons of two integers are required. The $O(r_s^{max})$ cost of checking if $n_i \notin V$ is of similar or inferior order to the $O(p)$ cost of converting SRV_S to an integer. Hence, for an aliasing test set, $X = p$, and the overall complexity of Algorithm 5.1 becomes $O(r_s^{max} \times N_{r_s} \times N_b \times p)$.

In structural diagnosis, the computation of T can be simplified in a similar fashion. In the case of structural diagnostic analysis with behaviorally synthesized aliasing-free STVs, T does not need to be computed (because it corresponds to the subgraph of shorted nets S) and no behaviors need be simulated. In this case, $X = 1$, and the complexity of diagnosis simulation reduces to $O(r_s^{max} \times N_{r_s})$. For an aliasing test set, SRV_S may correspond to a color that already exists in the set of colors CS. Testing this condition has complexity $O(p)$, as it can be shown. To compute T, the neighbors of each net with an erroneous response SRV_S are checked to verify if they also respond with SRV_S; this can be done in time $O(r_s^{max} \times D)$. In this way, if $p > r_s^{max} \times D$, then $X = p$; otherwise, $X = r_s^{max} \times D$. Hence, the overall complexity of diag-

[6] According to Section 3.3, for the most common sequences, such as the counting sequence plus complement scheme, p is in the order of $\log N$. For high diagnostic capability sequences, such as the walking sequences, p is in the order of N.

[7] To be rigorous, r_s^{max} does increase slightly with the circuit size, as discussed apropos of Equation 5.40.

[8] Due to STV uniqueness, in behavioral diagnosis there is at most one aliased net. For example, in the counting sequence the test vector of a net n_i is given by $STV_i = (i-1)_2$, i.e., STV_i is the p-bit binary representation of $i-1$. Then, if there is an aliasing short response SRV_S, the aliased net is a net n_j such that $j = SRV_S + 1$.

nosis simulation for aliasing structural schemes is $O(r_s^{max} \times N_{r_s} \times N_b \times p)$ if $p > r_s^{max} \times D$ and $O((r_s^{max})^2 \times N_{r_s} \times N_b \times D)$ if $p \leq r_s^{max} \times D$.

The short behaviors used in simulation are determined as follows. If a ground/power net is involved in the short, then only the S-A-0/1 behavior is considered, and the number of behaviors N_b is set to 1. Otherwise, the behaviors W-A, W-O and SD-k are considered. The W-A and W-O behaviors account for 2 behaviors and the S-D-k behaviors account for $\#V$ behaviors. [9] Thence, the number of behaviors is $N_b = \#V + 2$.

5.4 Summary

In this chapter a methodology for quantitative evaluation of diagnostic resolution has been presented. With such methodology, any two interconnect diagnosis schemes can be compared prior to their application, regardless of their type, behavioral or structural.

A method for actually evaluating DR in a manufacturing environment has been discussed. This method is based on equating DR to the average fault isolation efficiency. Some important empirical facts about behavioral interconnect diagnosis schemes have been analyzed. These facts are used to make the assumptions on which the DR assessment method is based.

The well known aliasing/confounding framework has been put in perspective. This method can be used to compare a behavioral scheme to another behavioral scheme in terms of diagnostic capability. The assessment provided is of qualitative nature, but the schemes can effectively be compared in relative terms using the method. Structural schemes can also be compared among themselves using the aliasing-confounding framework. However, it is incorrect to compare a behavioral to a structural scheme using the aliasing/confounding framework.

A previous attempt to develop a quantitative DR assessment method based on the concept of stain diagnosability has been reviewed. This approach is capable of correctly ranking structural schemes in terms of diagnostic capability. However, its main weakness resides in the fact that it does not have any useful physical meaning. Also, the stain-based approach only applies to structural diagnosis and can only be used to analyze short faults.

A new approach to DR assessment based on a concept of fault diagnosability has been presented. The fault-based approach expresses DR as the weighted average individual fault diagnosability. The individual fault diagnosability is defined as the ratio of the number of actual single faults by the number of diagnosed single faults. Fault statistics have been used to weight each fault according to its probability of occurrence. It is conjectured that the number of diagnosed single faults is proportional to the fault isolation time, so that DR is

[9] In a short $S = (V, E)$ there are as many SD-k behaviors as the number of possible strong-driver nets. Since each net in S is a potential strong-driver, then there are $\#V$ possible SD-k behaviors.

a measure of fault isolation efficiency in the presence of ambiguous diagnostic information.

Interconnect diagnosis simulation is used to compute DR before the test set under evaluation is applied. Since it is impractical to simulate the whole fault population, fault sampling criteria have been developed. Reasonable assumptions are made to obtain a manageable and yet representative fault sample. It has been concluded that open faults do not need to be sampled due to their often unambiguous diagnosability. Nonetheless, the contribution (γ_o) they give to DR is accounted for in the final DR score in order to avoid excess conservatism. The multiplicity and topology of the sampled shorts are constrained to avoid simulating faults with an insignificant occurrence rate. For each subsample of multiplicity r_s shorts, the average diagnosability d^{r_s} is computed up to a maximum multiplicity r_s^{max}. The shorts diagnostic resolution DR_s is computed by weighting each diagnosability d^{r_s} with a multiplicity probability $P_s(r_s|r_s > 0)$ obtained from the process characterization statistics. The overall DR is a function of γ_o and DR_s.

A diagnosis simulator that embodies the fault-based method for DR assessment has been developed. The complexity of the simulation algorithm varies according to type of scheme, behavioral or structural. For a behavioral scheme the complexity is $O(r_s^{max} \times N_{r_s} \times N_b \times p \times N)$ and for a structural scheme it is $O(r_s^{max} \times N_{r_s} \times N_b \times p \times r_s^{max} \times D)$. The diagnosis simulation algorithm may become inefficient for large EAs with long STVs. For these cases, several improvements have been suggested that effectively reduce the complexity of the algorithm.

6 EXPERIMENTAL RESULTS

In this chapter, experimental results on the various interconnect diagnosis methods are presented. The results illustrate fault extraction, diagnostic synthesis and diagnostic resolution assessment for a set of example EAs.

The EAs are assumed to be manufactured in the same process line, so that the same fault statistics apply to all of them. The statistical parameters have been given typical values found in the literature. The examples range from small designs to an extremely large design, so that test vector length, diagnostic resolution and algorithm performance can be studied as a function of the circuit size.

The chapter is organized as follows. In Section 6.1, the experimental environment is described. In Section 6.2, fault extraction results are presented. In Section 6.3, diagnostic synthesis and diagnosis simulation results are given. Section 6.4 presents a summary of the chapter.

6.1 Experimental setup

One of the main objectives of this work is to develop low complexity algorithms that can be run on modest computing facilities, enabling the use of low cost ATE systems. Following this philosophy, the algorithms have been run on a Sun Sparc 4M workstation, at 85 MHz, using 32 mega bytes of RAM and running the SunOS 4.1.4 operating system. The code has been written in the C language (Kernighan and Ritchie, 1978).

Using these computational resources, experiments on six example EAs have been carried out. The example EAs are PCB designs whose relevant characteristics are shown in Table 6.1. For each PCB the following characteristics are given: the number of nets N, the number of pins N_P and the ratio N_P/N.

Table 6.1. Board examples.

Board	N	N_P	N_P/N
ir232	22	80	3.64
68hc11	93	332	3.57
sram8	120	680	5.67
cperi24	271	952	3.51
hyp1k	1000	*4010*	*4.01*
hyp50k	50000	*200500*	*4.01*

The examples `ir232`, `68hc11`, `sram8` and `cperi24` are actual PCB designs for which the pin spacing PS is 100 *mil*. These circuits have been designed with the `Pcb` design package (Section 4.1.3). As was assumed in Section 4.1, the results for the four actual examples confirm that the ratio N_P/N does not necessarily increase with the circuit size.

The large `hyp1k` and `hyp50k` examples have been artificially generated. They are used to demonstrate the advantages of structural schemes and to test the performance of the algorithms for large circuit sizes. To test all but the fault extraction algorithm, all is needed is the adjacency graph G and some process statistical parameters. Hence, `hyp1k` and `hyp50k` simply consist of two randomly generated adjacency graphs whose structure, apart from the size, is similar to the structure of the graphs of the actual circuits. With its 50 thousand nets, `hyp50k` provides an extreme test case, as it is much larger than the average board. Many of the implemented computer programs do not complete for `hyp50k` after running for several days. However, when it is not possible to complete some program on `hyp50k`, its CPU time is estimated using the expression of the algorithm's complexity and the CPU time of `hyp1k`. This technique is also used to estimate the CPU time of algorithms that have not yet been implemented, but whose complexity is known. The performance of improvements suggested for various algorithms are evaluated in this way.

Since `hyp1k` and `hyp50k` are artificially generated graphs, their N_P values are not available. Nonetheless, it is important to know N_P in order to estimate the average number of faults λ. Based on the observation that N_P/N does not necessarily increase with the circuit size, it has been assumed that for the artificial examples $N_P/N = 4.01$, which is the average N_P/N for the four actual designs. Then, N_P has been computed as $N_P = 4.01 \times N$. The values of N_P and N_P/N for `hyp1k` and `hyp50k` appear in italics in Table 6.1 to emphasize the fact that they have been obtained in this way.

6.2 Fault extraction

A fault extraction program was implemented using Algorithm 4.1 described in Section 4.1. Assuming that the example EAs are fabricated in the same process line, and that the technology uses solder resist on both sides of the board, the fault extraction program has been run in the pin-to-pin mode. The maximum short radius has been set to $R = 1.5PS$. Recall that in the pin-to-pin mode R is normally set in the range $PS < R < 2PS$.

The fault extraction program produces the adjacency graph G for each input EA. Table 6.2 gives the relevant characteristics of the graphs obtained for each example: the number of vertices N, the number of edges M and the average vertex degree $D = 2 \times M/N$. For the first four examples (actual designs), it can be observed that D does not necessarily increase with the circuit size. Based on this observation, the `hyp1k` and `hyp50k` adjacency graphs have been generated while forcing D to have a value close to the average D of the four actual designs.

Table 6.2. Fault extraction results.

Board	N	M	D
ir232	22	40	3.33
68hc11	93	327	7.03
sram8	120	291	4.85
cperi24	271	1045	7.71
hyp1k	2120	6308	5.95
hyp50k	50000	143250	5.70

For the first four examples, the fault extraction program takes only a few seconds to run. For `hyp1k` and `hyp50k`, it is possible to estimate the time the fault extraction program would take to extract their artificially generated graphs from hypothetical layouts. For that purpose, the complexity expression of the fault extraction algorithm and the CPU time taken for one of the actual examples is used. The CPU time of example `cperi24` (4.38 seconds) is used for this purpose because it is the largest of the actual examples, and therefore its CPU time is less susceptible to be influenced by other factors. Since the complexity of the algorithm is $O(N^2)$, a CPU time of about 1 minute is estimated for `hyp1k` and a CPU time of about 2 days is estimated for `hyp50k`. The CPU time of `hyp1k` is reasonable, but 2 days for `hyp50k` is somewhat awkward. However, with the improvements suggested for the fault extraction algorithm in Section 4.1.3, its complexity could be reduced to $O(N \log N)$. With the improved algorithm, it is estimated that `hyp50k` would take only about half a minute to run.

6.3 Diagnosis

In this section, results on diagnostic synthesis and diagnosis simulation are presented. Diagnostic synthesis produces the test vector set and consequently the STV length p. Behavioral and structural diagnosis methods are analyzed using a statistical characterization of the manufacturing process. The diagnosis simulation results are explained by looking at the relationship between short diagnosability and short multiplicity.

6.3.1 *Statistical process characterization*

The statistical characterization of the manufacturing process is based on the results published for a Hewlett Packard SMT process, hereafter referred to as the HP process (Tegethoff, 1994a). In this process, the fault multiplicity distribution $P(r)$ is modeled using a negative binomial distribution with parameters λ and α (Equation (5.11)). The parameter λ has been estimated using Equation (5.18) for a solder fault rate $SFR = 50$ ppm. [1] The parameter α has been estimated using Equation (5.19); it has been empirically verified that $\alpha = 0.4$ characterizes the process for a range of board sizes. Small deviations from this value have been found uncorrelated with the circuit size.

In order to obtain a realistic process characterization, it has been decided to use the fault multiplicity distribution of the HP process: $SFR = 50$ *ppm* and $\alpha = 0.4$. However, to characterize the process according to the theory of Section 5.3.2, it is necessary to characterize the short and open fault multiplicity distributions separately. The shorts multiplicity distribution $P_s(r_s)$ is characterized by parameters SSR and α_s and the opens multiplicity distribution $P_o(r_o)$ is characterized by parameters SOR and α_o. The global fault multiplicity distribution $P(r)$ can be given in terms of the distributions $P_s(r_s)$ and $P_s(r_o)$ by the expression

$$P(r) = \sum_{r_s=0}^{r} P_s(r_s)P_s(r_o), \; r = r_s + r_o, \tag{6.1}$$

which, according to Equation (5.20) and Equation (5.21), is equivalent to

$$P(r) = \sum_{r_s=0}^{r} \frac{\Gamma(\alpha_s + r_s)}{r_s!\,\Gamma(\alpha_s)} \frac{(\frac{\lambda_s}{\alpha_s})^r}{(1+\frac{\lambda_s}{\alpha_s})^{r+\alpha_s}} \cdot \frac{\Gamma(\alpha_o + r_o)}{r_o!\,\Gamma(\alpha_o)} \frac{(\frac{\lambda_o}{\alpha_o})^r}{(1+\frac{\lambda_o}{\alpha_o})^{r+\alpha_o}}. \tag{6.2}$$

Since the HP process does not give separate information for open and short faults, the following assumptions have been made:

$$SSR = SOR = SFR/2 \tag{6.3}$$

and

$$\alpha_s = \alpha_o = \alpha/2. \tag{6.4}$$

[1] A solder fault rate of 50 ppm (parts per million), i.e., 50×10^{-6}, is the level of quality normally required for an industrial soldering process.

These assumptions mean that opens and shorts are equally important in distribution $P(r)$. It can be shown that Equation (5.11) can be obtained by substituting Equation (6.3) and Equation (6.4) in Equation (6.2). Thus, by making $SSR = SOR = 25$ *ppm* and $\alpha_s = \alpha_o = 0.2$, the process becomes statistically equivalent to the HP process. According to Section 5.3.3, to restrict the maximum short multiplicity that need be considered, a maximum error ϵ is imposed on the DR estimate. In the experiments reported here, ϵ has been set to 10^{-3}.

Table 6.3 gives the relevant statistical variables for each example, obtained for the process characterization just described. The statistical variables of interest are the average number of faults (λ), the overall yield (Y), the shorts yield (Y_s), the probability that only opens are present in a faulty EA (γ_o), the maximum DR_s error (ϵ_s) and the maximum short multiplicity (r_s^{max}). The value of λ is obtained from Equation (5.18), Y is given by Equation (5.12), Y_s is obtained from Equation (5.22), γ_o from Equation (5.36), ϵ_s from Equation (5.38) and r_s^{max} from Equation (5.40).

Table 6.3. Statistical process characterization.

Board	λ	Y	Y_s	γ_o	ϵ_s	r_s^{max}
`ir232`	0.00481	0.99523	0.99761	0.49940	0.00200	2
`68hc11`	0.01865	0.98194	0.99093	0.49772	0.00199	2
`sram8`	0.02406	0.97691	0.98839	0.49708	0.00199	2
`cperi24`	0.05434	0.95033	0.97485	0.49363	0.00198	3
`hyp1k`	0.20050	0.85000	0.92195	0.47970	0.00192	5
`hyp50k`	10.02500	0.27139	0.52095	0.34252	0.00152	112

Variable λ increases proportionally with the circuit size, which causes Y_s and Y to decrease. It may seem that the value of γ_o is too high. However, that is the value obtained assuming that the occurrences of shorts and opens are statistically independent, and that both multiplicity distributions are negative binomial. The probability that there are only short faults given the EA is faulty, which can be called γ_s, is given by $\gamma_s = \gamma_o$, since the statistics are symmetric for opens and shorts. This shows that the probability of mixed open-short faults is low when compared to γ_o or γ_s. For example, for `cperi24` the probability of mixed open-short faults is $1 - \gamma_o - \gamma_s = 0.01274$. As the circuit size increases, λ increases and, since α is kept constant, γ_s and γ_o decrease. The decline of γ_s and γ_o with the circuit size causes the probability of mixed open-short faults to increase. The error ϵ_s also decreases with the circuit size in order to keep the total error ϵ from increasing, [2] which forces r_s^{max} to increase (Equation (5.40)). In other words, as the circuit size increases, higher multiplicity shorts become more likely and need to be taken into account to keep DR high. The value

[2] In fact, according to Equation (5.38), ϵ_s converges to ϵ as γ_o decreases.

$r_s^{max} = 112$ for hyp50k seems somewhat large, which raises questions about the constancy of α at very large circuit sizes. Perhaps, for circuits of the size of hyp50k, α does increase with the circuit size to prevent r_s^{max} from being so high. Additionally, even if fault multiplicities in the order of 100 are likely in such large circuits, it is questionable whether connected short subgraphs of that size need to be analyzed. This issue is left open for further investigations.

A final aspect concerning the statistical process characterization is the distribution of fault behaviors $P(b)$ (Section 5.3.5). Due to the lack of actual statistical data, the distribution $P(b)$ has been assumed uniform for all fault behaviors: $P(b) = 1/3$ for b=W-A, $P(b) = 1/3$ for b=W-0 and $P(b) = 1/(3 \times \#V)$ for b=SD-k. Note that, since there are $\#V$ possible strong drivers in a short involving a set of nets V, the probability of strong-driver-logic behavior is also 1/3.

6.3.2 *Behavioral methods*

In this section, experimental results on behavioral diagnosis methods are presented and discussed. Behavioral interconnect diagnosis is the fault diagnosis method commonly used in the industry (Chapter 3). Three STV sequences have been selected to illustrate the behavioral method: the modified counting sequence, which has detection capability (Section 3.3.1); the self-diagnosis sequence, which has aliasing-free capability (Section 3.3.5); and the new N sequence, which has aliasing-free and confounding-free capabilities (Section 3.3.8). All these methods are symmetric with respect to W-A and W-O shorts, since W-A and W-O shorts have been assumed equally important.

Table 6.4. Results for behavioral diagnosis.

Board	Mod. count. seq.			Self-diag. seq.			N sequence		
	p	DR_s	DR	p	DR_s	DR	p	DR_s	DR
ir232	5	88.0	93.9	7	97.9	98.9	22	98.0	99.0
68hc11	7	85.2	92.5	9	99.0	99.5	93	98.7	99.3
sram8	7	74.7	87.2	9	98.0	99.0	120	98.7	99.3
cperi24	9	79.2	89.5	11	97.3	98.6	271	97.4	98.6
hyp1k	10	73.5	86.2	13	91.8	95.7	1000	91.8	95.7
hyp50k	16	39.3	60.1	19	44.3	63.4	50000	*44.3*	*63.4*

The experimental results are shown in Table 6.4. For each example and each STV sequence, the following results are given: the STV length p, the shorts diagnostic resolution DR_s and the overall diagnostic resolution DR. DR and DR_s are related by Equation (5.35). DR is always greater than DR_s because it includes the contribution of open faults, which always have diagnosability 1. According to the theory presented in Section 3.3, the STV length p of the modified counting sequence and the self-diagnosis sequence increases logarithmically with N; for the N sequence, p increases linearly with N. For the three sequences, DR declines with increasing circuit size. Note that higher multiplicity shorts become more likely as the circuit size increases, and that the individual fault diagnosabilities decrease with the fault multiplicity. [3]

The N sequence is aliasing-free and confounding-free, whereas the self-diagnosis sequence is simply aliasing-free. Since confounding shorts are not included in the fault sample, the two sequences have practically the same DR value. The small differences between them can be attributed to the use of distinct fault samples (Section 5.3.4), and tend to reduce with the increase of the sample size N_{r_s}. Comparing the DR columns of the two sequences in Table 6.4, it can be observed that the fluctuations never exceed 1%, which indicates that $N_{r_s} = 1000$ is an adequate sample size. As discussed in Section 5.3.3, it is possible that the exclusion of confounding shorts in behavioral diagnosis favors the self-diagnosis sequence in terms of DR.

The DR difference between the self-diagnosis and the modified counting sequence tends to decrease with the circuit size, being of only about 4% for `hyp50k`. As high multiplicity shorts become more likely with the circuit size, the difference of individual fault diagnosabilities between aliasing and aliasing-free test sets declines. In fact, consider the following: first, the set of suspect nets U given by an aliasing-free test set has at most one net less than the set of suspect nets given by an aliasing test set; second, the number of suspect single shorts r'_s increases quadratically with the number of suspect nets $\#U$ (Equation (5.47)); third and last, the individual fault diagnosabilities vary inversely with r'_s (Equation (5.46)). Then, by means of a simple calculation, it can be concluded that the DR difference between aliasing and aliasing-free schemes decays on average with $1/(\#U)^3$. It does not mean that aliasing-free schemes are unnecessary for large circuit sizes. Note that the yield Y also decreases with the circuit size, causing the costs with diagnosis, fault isolation, repair and retest to increase. When Y is low, the term $(1 - Y)(CDI + CRR)$ in Equation (1.1) can become a considerable fraction of the total unitary cost CU. Hence, it is important to have a high DR to keep the unitary cost of diagnosis and fault isolation (CDI) as low as possible. Note that DR and CDI are inversely related, as discussed in Section 5.1.1.

The diagnosis simulation program runs in a matter of seconds for the first four examples. For `hyp1k` with the modified counting sequence and the self-diagnosis sequence, it also takes a few seconds. For the N sequence, `hyp1k` takes 42 minutes. Example `hyp50k` takes 12 hours and 32 minutes for the modified counting sequence, and 14 hours and 53 minutes for the self-diagnosis sequence. For the N sequence the `hyp50k` simulation was aborted. According to the $O(r_s^{max} \times N_{r_s} \times N_b \times N \times p)$ complexity of the diagnosis simulation algorithm in the behavioral mode, it is estimated that this simulation would run for about 5 years.

[3] High multiplicity connected shorts cause diagnostic algorithms to produce larger sets of suspect nets U. Since the number of suspect single shorts r'_s increases with $(\#U)^2$ (Equation (5.47)), the individual short diagnosabilities decrease with $1/(\#U)^2$.

DR_s and DR results for `hyp50k` with the N sequence have been obtained by weighting the diagnosabilities d^{r_s} of `hyp1k` with the fault multiplicity distribution of `hyp50k`. These results appear in italics in Table 6.4 to signal the fact that they have been obtained in this way. It is being assumed that the d^{r_s} values for `hyp1k` and `hyp50k` are similar, since their adjacency graphs have a similar structure. Because faults of multiplicity up to $r_s^{max} = 112$ must be considered, the simulation takes almost one day. However, with the improvements suggested in Section 5.3.6, the complexity of diagnosis simulation in the behavioral mode can become $O(r_s^{max} \times N_{r_s} \times N_b \times p)$ for aliasing sequences, or $O(r_s^{max} \times N_{r_s} \times N_b)$ for aliasing-free sequences. Thus, with the improved algorithm, `hyp50k` would run in about 43 minutes for the modified counting sequence, and in less than a second for the self-diagnosis and the N sequences.

6.3.3 Structural methods

In this section, results on structural diagnosis methods are presented. Structural methods with the following STV types are analyzed: behavioral, graph coloring and statistical color mixing STVs. The same STV sequences used in the previous section are considered: the modified counting sequence, the self-diagnosis sequence and the N sequence.

Behavioral STVs. In structural diagnosis with behavioral STVs, test sets are synthesized using a behavioral method and diagnostic analysis is performed using a structural method. This method is described in Section 4.4.1, and some experimental results are presented in Table 6.5.

Table 6.5. Results for structural diagnosis with behavioral STVs.

Board	Mod. count. seq.			Self-diag. seq.			N sequence		
	p	DR_s	DR	p	DR_s	DR	p	DR_s	DR
`ir232`	5	98.7	99.3	7	99.1	99.5	22	99.9	99.9
`68hc11`	7	98.0	98.9	9	99.7	99.8	93	99.5	99.7
`sram8`	7	95.1	97.5	9	99.8	99.9	120	99.8	99.9
`cperi24`	9	96.3	98.1	11	99.3	99.6	271	99.4	99.7
`hyp1k`	10	99.3	99.6	13	99.8	99.9	1000	99.8	99.9
`hyp50k`	16	88.9	92.7	19	89.0	92.7	50000	*89.0*	*92.8*

Test sets for this method are the same as in the previous section, and thus the STV lengths p of Table 6.5 equal the STV lengths of Table 6.4. In contrast, the diagnostic resolution of this method is better than the diagnostic resolution of the behavioral method. Structural diagnostic analysis is so powerful that, despite their different diagnostic capabilities, the schemes can barely be distinguished in terms of DR. Note that the modified counting sequence is only slightly worse than the self-diagnosis and the N sequences, which indicates that the aliasing probability is very small for this method. Structural diagnostic

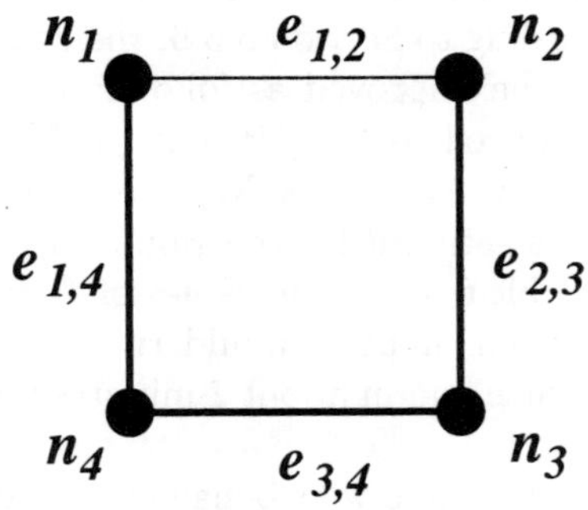

Figure 6.1. Intrinsic diagnostic ambiguity.

analysis resolves many situations that would be ambiguous in behavioral diagnosis, and the use of behavioral STVs ensures that the diagnostic capabilities do not degrade with the short multiplicity. This is the best possible diagnostic method in terms of DR and will be used as a reference for the maximum achievable DR.

Since only aliasing shorts are being considered, it is logical that both the self-diagnosis and the N sequence show a similar DR value. Nonetheless, even if confounding shorts were to be considered, the N sequence would not do much better than the self-diagnosis sequence. In fact, only adjacent short subgraphs can confound when structural diagnostic analysis is being performed, which is deemed unlikely.

Given that both the self-diagnosis and the N sequences are aliasing-free, one may ask why DR is not 1 for these sequences. The answer can be given using the concept of *intrinsic diagnostic ambiguity.* In the graph shown in Figure 6.1, the short $S = (V, E'')$, where $V = \{n_1, n_2, n_3, n_4\}$ and $E'' = \{e_{1,4}, e_{3,4}, e_{2,3}\}$, is considered. For this short, the stain $T = (V, E')$, where $E' = \{e_{1,4}, e_{3,4}, e_{2,3}, e_{1,2}\}$, is computed. It can be seen that no nets are aliased. Nonetheless, the diagnosability of short S is $\#E''/\#E' = 3/4 < 1$. This example shows that diagnostic ambiguity is possible even without aliasing and confounding. The ambiguity is due to the topology of the stain and it is only possible in cyclic stain subgraphs. Since this type of diagnostic ambiguity cannot be avoided by using a different test set, it has been called *intrinsic diagnostic ambiguity.*

The diagnosis simulation program completes in a few seconds for the first four examples. It also takes a few seconds for `hyp1k` with the modified counting sequence and the self-diagnosis sequence. For the N sequence, `hyp1k` takes about 4 minutes. Example `hyp50k` takes about 34 minutes for the modified counting sequence and 40 minutes for the self-diagnosis sequence. For the N sequence, the `hyp50k` simulation was aborted. Since the complexity of diagnosis simulation in the structural mode is $O((r_s^{max})^2 \times N_{r_s} \times N_b \times p \times D)$, it is estimated that this simulation would take about 70 days to run. The DR_s and DR results have again been obtained by simulating `hyp1k` using the short multiplicity distribution of hyp50k. For this reason, these results appear in italics in Table 6.5. According to Section 5.3.6, the complexity of the diagnosis simulation algorithm can be improved as follows: for aliasing-free sequences, the complexity can be reduced to $O(r_s^{max} \times N_{r_s})$; for aliasing sequences, the complexity can be reduced to $O(r_s^{max} \times N_{r_s} \times N_b \times p)$ if $p > r_s^{max} \times D$. The condition $p > r_s^{max} \times D$ is satisfied for the values of p given in Table 6.5, the values of r_s^{max} given in Table 6.3 and the values of D given in Table 6.2. Thus, with the improved algorithm, hyp50k would run in 0.2 seconds for the self-diagnosis and the N sequence and in about 2 minutes for the modified counting sequence.

The structural diagnosis method with behavioral STVs is a good choice when the STV length produced by the behavioral schemes is acceptable. The method is useful in a situation where a large EA is being produced in high volumes. For a large EA the yield is low and a high diagnostic resolution is needed to minimize diagnosis and fault isolation time. To store and apply the behavioral STVs, an ATE system with enough capacity is required. The price of such a system may be high, but it can be justified by the high production volumes involved.

Graph coloring STVs. Structural diagnosis with graph coloring STVs yields the shortest possible serial test vectors but has the worst diagnostic capabilities in terms of aliasing and confounding. The graph coloring method is given by Algorithm 4.3 of Section 4.4.2. The experimental results are given in Table 6.6. Note that in structural diagnostic synthesis the N sequence is renamed the K sequence.

Table 6.6. Results for structural diagnosis with graph coloring STVs.

Board	Mod count. seq.			Self-diag. seq.			K sequence		
	p	DR_s	DR	p	DR_s	DR	p	DR_s	DR
ir232	3	83.3	91.6	3	87.5	93.7	3	82.3	91.1
68hc11	3	82.4	91.1	4	88.1	94.0	5	78.1	89.0
sram8	3	65.3	82.5	4	77.6	88.7	4	62.7	81.2
cperi24	3	78.3	89.0	4	86.2	93.0	6	75.5	87.6
hyp1k	3	79.6	89.3	4	87.0	93.2	4	78.5	88.8
hyp50k	*3*	*82.5*	*88.5*	*4*	*84.8*	*90.0*	*4*	*82.5*	*88.5*

The STV lengths p are lower than the STV lengths of the behavioral STVs (Table 6.4), and do not increase with the circuit size. In terms of DR, the method is worse than the behavioral diagnosis method for all circuits except hyp50k. In fact, in behavioral diagnosis the individual fault diagnosabilities decay faster with the circuit size than in structural diagnosis with graph coloring vectors. [4] Compared with structural diagnosis with behavioral STVs (Table 6.5), which sets the maximum achievable DR, the graph coloring method has a considerably inferior DR. The DR difference between the two methods tends to reduce with the increase in circuit size, but, as mentioned in Section 6.3.2, the importance of DR also increases with the circuit size due to the decline of Y.

For the K sequence DR is worse than for the modified counting sequence. This confirms the theory presented in Section 4.4.2 according to which the global diagnosability property of the N sequence is not applicable to the K sequence. The results show that not only global diagnosability works poorly with graph coloring vectors, as it makes the K sequence even more ambiguous than the modified counting sequence.

The complexity of Algorithm 4.3 is determined by the complexity of its graph coloring step, which is performed by Algorithm B.1 described in Appendix B. For the first three examples, the graph coloring CPU time is on the order of a few seconds. For `cperi24` and `hyp1k` it takes about 9 seconds and 10 minutes, respectively. The graph coloring program could not be completed for `hyp50k`, for which a CPU time of about $2\frac{1}{2}$ years is estimated for the current $O(N^3)$ version of the program. With the improvements suggested for the algorithm, its complexity can become $O(D \times N)$, which means that `hyp50k` would run in about 0.2 seconds. Since the chromatic number of a graph has more to do with the structure of the graph than with its size, and since the `hyp1k` and `hyp50k` graphs have a similar structure, it has been assumed that `hyp50k` can be colored with the same number of colors as `hyp1k`. Thus, `hyp50k` is assumed to produce the same STV length as `hyp1k`; the p results appear in italics in Table 6.6 to signal the fictitious way in which they have been obtained.

The diagnosis simulation program completed in a few seconds for the first five examples. Since it has not been possible to color graph `hyp50k`, diagnosis simulation was carried out on `hyp1k`, using the short multiplicity distribution of `hyp50k`. The program took about 1 hour for the self-diagnosis and the N sequences, and about 44 minutes for the modified counting sequence. The reason why it takes so long is the fact that the maximum short multiplicity $r_s^{max} = 112$ is high when compared to the other examples (Table 6.3). Note that the complexity $O((r_s^{max})^2 \times N_{r_s} \times N_b \times p \times D)$ of diagnosis simulation in the structural mode depends quadratically on r_s^{max}. In Section 5.3.6, it has been shown that the complexity of diagnosis simulation can be improved to $O((r_s^{max})^2 \times N_{r_s} \times N_b \times D)$ if $p \leq r_s^{max} \times D$. The condition $p \leq r_s^{max} \times D$ is satisfied for the values of p given in Table 6.6, the values of r_s^{max} given in Table 6.3 and the values of D given in Table 6.2. Thus, with the improved version of the program, the CPU time for `hyp50k` is estimated to be about 15 minutes, for any STV sequence.

---

[4] In behavioral diagnosis the individual fault diagnosabilities vary as $1/(\#U)^2$, whereas in structural diagnosis it can be observed that the number of aliased shorts in a stain decreases at a much slower pace.

Structural diagnosis with graph coloring STVs is useful in situations where the main issue is test data size, DR being less important. Such situation occurs when, for example, a small EA is being produced in small volume. A small EA has a high yield, which means that the total number of EAs to be diagnosed and repaired is low. With a low production volume, the STV length should be as short as possible to reduce the capacity and cost of the ATE system. Structural diagnosis with graph coloring STVs enables just that: the lowest possible STV length. The low DR of the method should not matter because, the yield being high, spending more time to isolate faults in few faulty EAs will not significantly impact the final cost.

Statistical color mixing STVs. Structural diagnosis with statistical color mixing STVs produces serial vectors slightly longer than the graph coloring method but gives a substantially higher diagnostic capability. The working principle of this method is based on the concept of *extent* (Definition 4.16): all nets in a stain whose extent is lower than a certain maximum stain extent ε are assigned unique STVs. The statistical color mixing method is given by Algorithm 4.5 of Section 4.4.4. Experimental results for $\varepsilon = 2$ are shown in Table 6.7.

Table 6.7. Results for structural diagnosis with statistical color mixing STVs.

Board	Mod count. seq.			Self-diag. seq.			K sequence		
	p	DR_s	DR	p	DR_s	DR	p	DR_s	DR
ir232	3	93.0	96.5	4	99.9	99.9	6	97.8	98.9
68hc11	4	93.9	96.9	6	99.5	99.7	14	99.7	99.8
sram8	4	89.8	94.8	6	99.7	99.8	13	99.6	99.8
cperi24	5	93.6	96.7	7	99.2	99.6	29	99.2	99.5
hyp1k	5	95.0	97.4	6	99.5	99.7	17	99.4	99.7
hyp50k	*5*	*87.3*	*91.7*	*6*	*88.8*	*92.6*	*17*	*88.7*	*92.6*

The STV length p, although higher than the STV length of the graph coloring method (Table 6.6), is still significantly lower than the STV length of behavioral STVs (Table 6.4). Similar to the graph coloring method, p does not tend to increase with the circuit size. The diagnostic resolution of the statistical color mixing method is almost as high as the diagnostic resolution of the structural method with behavioral STVs. The modified counting sequence has the worst DR but its value is still impressively high.

The complexity of Algorithm 4.5 is dominated by the $O(\varepsilon \times N^3)$ complexity of its graph enhancing step, which is performed by Algorithm C.1 described in Appendix C. For the first four examples the program takes a few seconds of CPU time; for hyp1k it takes about 17 minutes. The simulation of hyp50k was aborted, but it is estimated that it would run for about 4 years. With the improvements suggested for Algorithm C.1, its complexity can become $O(N \times$

D^{ε}), which would improve the graph enhancing CPU time of hyp50k to about 2 seconds.

The next step in Algorithm 4.5 is the coloring of the enhanced graph G_{ε}, which is performed by Algorithm B.1 described in Section B. CPU times for graph coloring are approximately the same as in the previous section, since the $O(N^3)$ complexity of the algorithm depends mainly on the number of vertices N. With the improvements suggested for the graph coloring algorithm, it would be possible to obtain a $O(D \times N)$ algorithm. The improved algorithm is sensitive to the vertex average degree D and therefore would have different runtimes for G and G_{ε}. Graph G would run D_2/D times faster than graph G_2. It has been observed that for all the examples the factor D_2/D is roughly 4. Hence, it is estimated that the improved graph coloring algorithm would take about 0.8 seconds to color G_2 for hyp50k.

Diagnosis simulation takes a few seconds for the first five examples. Since it has not been possible to perform graph enhancing on hyp50k, diagnosis simulation has been performed on example hyp1k using the short multiplicity distribution of hyp50k. The CPU time is about 1 hour 25 minutes for the N sequence, 30 minutes for the self-diagnosis sequence and about 23 minutes for the modified counting sequence. Considering that the complexity of diagnosis simulation in the structural mode is $O((r_s^{max})^2 \times N_{r_s} \times N_b \times p \times D)$, it can be observed that the CPU time per test vector is lower for the statistical color mixing schemes than it is for the graph coloring schemes. The explanation for this fact is the lower diagnostic ambiguity of the statistical color mixing schemes. If faults can be diagnosed with lower diagnostic ambiguity, the amount of processing per fault reduces, causing the overall CPU time to decrease. With the improvements suggested for the diagnosis simulation algorithm, its complexity can be reduced to $O((r_s^{max})^2 \times N_{r_s} \times N_b \times D)$, which would reduce the CPU time of hyp50k approximately to 5 minutes.

The structural method with statistical color mixing STVs proposed in this work can effectively minimize p while maximizing DR. The method is useful when manufacturing large EAs in medium production volumes. For a large EA the yield is low and consequently a high DR is required. For medium production volumes the cost of the ATE system should be mitigated by shortening the STV length p. Therefore, the structural method with statistical color mixing STVs is definitely a good option in a situation like this.

6.3.4 *Diagnosability versus short multiplicity*

The results of the previous two sections can be better understood by plotting the estimates $d^{r_s}{}'$ of the average diagnosability d^{r_s} as a function of the short multiplicity r_s for the diagnostic methods in question. The plots shown in Figure 6.2 have been obtained using hyp50k and the self-diagnosis sequence. Example hyp50k has been chosen because the multiplicity r_s spans over a large range, from $r_s = 1$ to $r_s = 112$. Figure 6.2 shows results up to $r_s = 20$ and also gives the probability $P_s(r_s|r_s > 0)$ of a multiplicity r_s short, given some

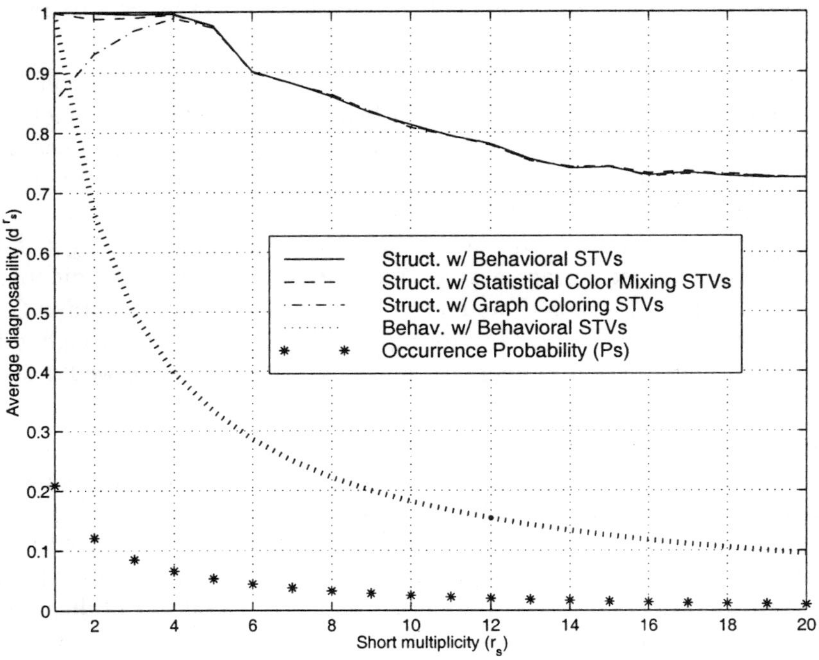

Figure 6.2. Diagnosability versus short multiplicity.

short has occurred. These probabilities represent the weight assigned to each diagnosability d^{r_s}.

For the behavioral method d^{r_s} is 1 for single shorts ($r_s = 1$) and decreases rapidly with r_s for $r_s > 1$. That is due to the decay of the individual fault diagnosabilities with $1/(\#U)^2$ (Equation (5.47)). The smaller values of $d^{r_s\prime}$ for this method explain the lower DR values obtained in Table 6.4.

Figure 6.2 shows that structural methods with behavioral STVs and statistical color mixing STVs produce the highest $d^{r_s\prime}$ estimates for any r_s. According to Section 6.3.3, the structural method with behavioral STVs is supposed to produce the maximum achievable diagnosabilities d^{r_s}. As can be seen, the structural method with statistical color mixing STVs is practically as effective as the structural method with behavioral STVs. The $d^{r_s\prime}$ estimates for the two methods are virtually indistinguishable, apart from some statistical fluctuations caused by fault sampling, which attenuate with the sample size. These fluctuations explain why the $d^{r_s\prime}$ estimates for the structural method with statistical color mixing STVs are, for some values of r_s, slightly above the $d^{r_s\prime}$ estimates for the structural method with behavioral STVs. For these methods $d^{r_s\prime}$ is practically 1 for short multiplicities r_s up to $r_s = 4$. Note that it is most important that d^{r_s} is high for low multiplicity faults, since those are the faults with highest probabilities of occurrence. It is exactly by achieving high diagnosability at low fault multiplicities that these methods accomplish the highest possible DR. For $r_s > 4$, d^{r_s} decays slowly with r_s due to intrinsic diagnostic ambiguity (Section 6.3.3). Note that, for $r_s \geq 16$, $d^{r_s\prime}$ seems to stabilize at about 0.72, suggesting that the intrinsic diagnostic ambiguity saturates with r_s.

For the structural method with graph coloring STVs, the $d^{r_s\prime}$ values are lower than those for the structural methods with behavioral and statistical color mixing STVs. The $d^{r_s\prime}$ value increases with r_s up to $r_s = 4$, where it merges with the other two curves in Figure 6.2. The fact that d^{r_s} is low for $r_s \leq 4$ causes the DR value of the graph coloring schemes to be significantly lower than the DR value of the other two methods. It seems strange that $d^{r_s\prime}$ increases with r_s in the interval $1 \leq r_s \leq 4$. Normally, the diagnosability decreases with fault multiplicity and not the other way round. After some investigations the reason has been discovered. As r_s increases, the probability of shorts to ground or power nets also increases due to the ubiquitousness of this kind of nets in the layout. Since these shorts give rise to S-A-0/1 behavior, and this behavior can be diagnosed with low ambiguity, it becomes clear why the $d^{r_s\prime}$ values increase. This effect partly explains why $\varepsilon = 2$ is enough to ensure an adequate DR. Compared to the graph coloring method the statistical color mixing method is better only at low short multiplicities; at higher multiplicities, the two methods are equally effective. Thus, $\varepsilon = 2$ is enough to improve the diagnosability at low short multiplicities. At higher multiplicities, the effect of ground and power nets determines the diagnosability. In circuits where there are not as many ground and power nets, ε may need to be set to a higher value. To test this hypothesis experiments with circuits that had ground and power nets excluded from the adjacency graph have been performed. For these circuits ε had to be set to 3 or 4 in order to achieve an adequate DR. These experiments also showed that even for $\varepsilon = 3$ or $\varepsilon = 4$, the STV length of the statistical color mixing method is still considerably lower than the STV length of the behavioral STVs. Additionally, the saturation of the intrinsic diagnostic ambiguity with r_s, makes it useless to increase ε above 3 or 4, even in the absence of ground and power nets. Both the ground/power nets and the intrinsic ambiguity effects contribute to stabilize the statistical color mixing method. Note that the main weakness of this method is having to set ε to a high value in order to achieve an adequate DR. Increasing ε may cause p to increase to a value where it no longer can compete with the behavioral STVs. [5] Fortunately, these effects and the rapid decay of the multiplicity probability make it unnecessary to increase ε beyond 3 or 4.

6.4 Summary

In this chapter experimental results on various interconnect diagnosis methods have been presented. The results illustrate fault extraction, diagnostic synthe-

[5] Increasing ε causes the chromatic number of G_ε to increase rapidly, which, in turn, causes p to increase (Section 4.4.4).

sis and diagnosis simulation, and have been obtained using a set of six PCB examples.

The fault extraction algorithm is shown to be effective, taking a few minutes for circuits of moderate sizes. For a very large example the program could take a few days, which may not be a problem since fault extraction is only performed once. If that were a problem, an improved version of the fault extraction algorithm suggested in Section 4.1 could be used instead. With this version of the algorithm, it is estimated that the largest boards would run in less than a minute.

The interconnect diagnosis methods have been simulated using a statistical process characterization based on the results published for an actual SMT process (Tegethoff, 1994a). The results clearly demonstrate the potential of structural interconnect diagnosis. Compared to the traditional behavioral diagnostic schemes, structural diagnosis can reduce the amount of test data while improving the diagnostic resolution. Three types of structural interconnect diagnosis methods have been considered: with behavioral, graph coloring and statistical color mixing STVs.

Structural diagnosis with behavioral STVs has been proposed, and is shown to provide the best possible diagnostic resolution. The combination of behavioral vectors with structural diagnostic analysis is powerful: many faults that would be ambiguous in behavioral diagnosis can effectively be disentangled by structural analysis, with no need to improve the test set. One example is the modified counting sequence scheme, which, despite being one of the most ambiguous schemes in behavioral diagnosis, produces a very competitive diagnostic resolution when combined with structural analysis. Also, the use of behavioral STVs whose diagnostic capabilities do not degrade with the fault multiplicity is another important reason for the effectiveness of the method.

Structural diagnosis with graph coloring STVs (Garey et al., 1976) can provide the shortest possible STV length at the cost of sacrificing some diagnostic resolution. In fact, its DR is worse than the DR of the traditional behavioral method for circuits of a medium size or lower. However, the method becomes as good or better than the behavioral method for very large circuits. The reason for that is the increased occurrence probability of higher multiplicity faults with the circuit size. Since the diagnosability of the graph coloring method decreases slower than the diagnosability of the behavioral method, for circuits above a certain size, the graph coloring method becomes better than the behavioral method.

Structural diagnosis with statistical color mixing STVs, as proposed here, was inspired by the previous schemes found in the literature (Lien and Breuer, 1991; McBean and Moore, 1993). With the new method both the STV length and the diagnostic resolution can be optimized. The STVs are slightly longer than the graph coloring STVs. However, the diagnostic resolution is practically as high as the diagnostic resolution of the structural method with behavioral STVs, i.e., it is very close to the maximum achievable DR.

It has been demonstrated that the computational cost associated with structural interconnect diagnosis is very affordable. All algorithms have been implemented and run on a modest computing facility: a Sun Sparc 4M workstation, at 85 MHz, using 32 mega bytes of RAM and running the SunOS 4.1.4 operating system. For the largest of the examples, which has intentionally been generated with an extreme size — 50 thousand nets — practically all the implemented algorithms ran into difficulties. For circuits of that size it has been demonstrated that the improvements suggested for the algorithms can effectively improve their performance. That has been accomplished by estimating the CPU time of the algorithms using their complexity expressions. Except for diagnosis simulation, which can take several minutes to run, the proposed improved versions can all run in a matter of seconds.

7 CONCLUSION

This book addresses the usage of boundary-scan (BS) for test and diagnosis of wire interconnect faults. The methods presented are applicable to any environment where wire interconnects and logic are separated. Applications other than BS include test and diagnosis of bare boards and interconnects of Field Programmable Gate Arrays (FPGAs). BS can also be used for testing non-compliant clusters of digital components or testing on-board chips individually. However, in modern electronic assemblies (EAs), non-compliant components are avoided and built-in self test is preferred for testing individual ICs in the EA. Among the various applications of BS, test and diagnosis of interconnect faults will remain significant.

In the last three decades, many interesting papers have been published on this subject. This book is a synthesis of those publications, including our own perspective and contributions to this field. Other books have focused on the BS standard itself, giving an in-depth description of the test hardware and embedded IC test. In this book, we concentrate on one of its most important applications.

This chapter recapitulates on the main topics of this work and gives directions for future work. The organization of the chapter follows the organization of the book itself: each section corresponds to a chapter of the book. Section 7.1 outlines the technological background introduced in Chapter 1. Section 7.2 summarizes the interconnect circuit and fault models of Chapter 2. Section 7.3 discusses the analysis of the behavioral diagnostic schemes presented in Chapter 3. In Section 7.4, the structural diagnosis methods studied in Chapter 4 are summarized. Section 7.5 outlines the diagnostic resolution assessment method of Chapter 5. The main experimental results presented in Chapter 6 are summarized in Section 7.6. Finally, Section 7.7 discusses possible future developments.

7.1 Technology context

Production defects always occur when complex electronic assemblies (EAs) such as PCBs, MCMs or SOCs are being manufactured. Among the production defects, interconnect defects caused during the assembly process are dominant. This justifies the importance of this work.

The cost of testing and diagnosing faulty EAs depends on two main variables: the test data size and the diagnostic resolution. The former influences the cost of ATE systems and test application time, whereas the latter impacts fault isolation time and repair cost.

Until recently, the most well known technology for EA test and diagnosis has been the in-circuit test (ICT) technology (Bateson, 1985). ICT relies on the ability to physically access the nets for applying and collecting test signals and responses. However, due to the increasing miniaturization of EAs, ICT is rapidly becoming obsolete.

Currently, the most common technology for testing and diagnosing complex EAs is the boundary-scan (BS) technology. The methods described in this work target this technology. In a BS chip, I/O pins are equipped with BS cells, which are chained together to form a large BS shift register. Serial test vectors (STVs) are applied into one end of the chain and serial response vectors (SRVs) are collected at the other end of the chain. The ICs themselves are chained together to form an EA level BS shift register, which provides access to each pin of each IC in the EA. BS is the IEEE 1149.1 standard and has been developed for digital ICs (IEEE, 1990; Parker, 1998). The methodology has been extended to system T&D (IEEE 1149.5) (IEEE, 1995), and recently to analog and mixed-signal ICs (IEEE 1149.4) (IEEE, 1999; Osseiran, 1999).

Due to the serial application of BS test vectors, it is useful to minimize the length of the STVs. Compact test sets save time in test application and reduce the cost of the ATE system. High diagnostic resolution is also key to reducing the cost of fault isolation. We have analyzed the trade offs between these two contradictory objectives.

7.2 Modeling the interconnect circuit and faults

Models for the interconnect circuit and interconnect faults have been presented. The interconnect circuit model assumes that each net has a driver end where the diagnostic stimuli is applied, and a receiver end where the diagnostic responses are collected. It has been shown that T&D schemes can be studied for single driver and single receiver nets, without loss of generality; it is straightforward to generalize the methods for nets with multiple drivers and/or multiple receivers.

Interconnect fault models have been categorized in *synthetic* and *analytic* models. The former are used to derive diagnostic input stimuli and the latter to analyze diagnostic responses. Two types of faults are considered: shorts and opens. The synthetic fault models for shorts among signal nets are the *wired-logic* and the *strong-driver-logic* models. The strong-driver-logic is a very common short behavior, which, in traditional behavioral methods, is subsumed by wired-logic models and does not need to be explicitly considered. However, we show that in structural diagnostic synthesis the strong-driver behavior should be explicitly considered. For shorts to ground/power nets and open faults, we have considered the *stuck-at-logic* model. As analytic fault models, we have considered the *same-response* and *any-response* models. With the former, any nets that produce the same response can be interpreted as potentially shorted; with the latter, any net that produces an erroneous or non-unique response can be diagnosed as potentially open. The stuck-at-logic fault model is also used as an analytic model to diagnose nets that produce either a sequence of 0s or a sequence of 1s. This behavior normally indicates that the net is potentially open or shorted to a ground/power net.

7.3 Behavioral interconnect diagnosis

Behavioral interconnect diagnosis is the method traditionally used to test and diagnose interconnect faults. In this book an original analysis of these methods is presented. In behavioral diagnosis, it is assumed that (1) any net can be open, and (2) any two nets can be shorted. Assumption (1) is generally true. However, (2) is a conservative assumption. In fact, only nets that are physically close can be shorted, but this information is not taken into account in behavioral diagnosis.

Using the analytic fault models, a $O(N \log N)$ diagnostic analysis procedure has been described, where N is the number of nets in the EA. The following classification of diagnoses is presented: a diagnosis can be complete, incomplete or incorrect; complete or incomplete diagnoses can be ambiguous or unambiguous; a complete and unambiguous diagnosis is called perfect. This classification enables distinguishing between different levels of diagnostic capability.

The diagnostic capability of various behavioral interconnect diagnosis schemes described in the literature has been studied in detail. It has been proven that any composite fault whose subcomponents behave according to the synthetic fault behaviors can be detected and diagnosed in a finite number of test-diagnose-repair cycles. The diagnostic capabilities of the schemes have been analyzed based on the well known aliasing/confounding framework. Also, a new diagnosis sequence, which has the property of being the shortest diagonally independent sequence, has been derived. Diagonal independence is a property that automatically guarantees aliasing-freedom for all fault behaviors, and confounding-freedom for almost all pairs of behaviors.

7.4 Structural interconnect diagnosis

In structural interconnect diagnosis, only the faults that are physically possible are taken into account for deriving test sets or analyzing test responses. Compared to traditional behavioral methods, these methods can produce more compact test sets and/or enable higher diagnostic resolution. STV length can be traded off for diagnostic resolution without ever exceeding the STV length of behavioral methods.

In structural diagnosis, layout fault extraction is needed to identify all layout locations where faults can potentially occur. A PCB layout fault extraction algorithm of complexity $O(N^2)$ has been described and implemented. It is explained how the algorithm can be improved for a $O(N \log N)$ complexity. The program extracts potential open faults at solder joints and potential shorts in two possible modes: between pins (pin-to-pin mode) or between pins and neighbor nets (pin-to-net mode). The output of the fault extraction program is a graph G called the adjacency graph. The vertices of G represent nets (or open faults), and its edges represent potential shorts between two nets.

Structural diagnostic analysis can be accomplished in $O(N)$ time and is a powerful tool for diagnosis, which, combined with an effective test set, can maximize the diagnostic capability. On the other hand, structural analysis allows more compact test sets to be used while still guaranteeing adequate diagnostic resolution. A structural diagnosis method that uses structural analysis but maintains the behavioral synthesis of test vectors has been proposed. With this method, outstanding diagnostic resolution can be achieved, since many of the diagnostic ambiguities are resolved by the structural information.

The graph coloring method for diagnostic synthesis has been analyzed in detail (Cheng et al., 1990). It has been derived from an earlier method for detecting shorts in PCB layouts (Garey et al., 1976). This method is based on coloring the vertices of the adjacency graph G, so that any neighbor nets are assigned different colors. Each color represents a distinct STV, allowing test vectors to be massively reused. The authors of the graph coloring method have claimed that its diagnostic properties are the same as those of the STV sequence assigned to the set of colors. The method can indeed produce short test sequences capable of fault detection but, unfortunately, the diagnostic capabilities do not match those of the behavioral sequence used. We have pointed out that in any method that allows test vector reuse, it is impossible to completely eliminate aliasing and confounding ambiguities.

The aliasing problem in structural diagnosis only occurs when strong-driver shorts are considered, and for this reason it has not been reported in the literature. However, the confounding problem also manifests itself with wired-logic shorts, and had already been noticed by other researchers who tried to solve it using color mixing methods (Lien and Breuer, 1991; McBean and Moore, 1993). We have observed that these methods can reduce strong-driver aliasing and all types of confounding, but can never eliminate them completely. Based on this observation, a statistical color mixing approach has been proposed. This approach can eliminate diagnostic ambiguities within a certain user specified subgraph extent, which is set according to statistics on fault multiplicity.

The proposed method is based on a technique called graph enhancing, which adds extra edges between any vertices whose edge-distance in graph G does not exceed a certain limit ε called the maximum extent. The diagnostic capability can be tuned by varying the value of ε: if $\varepsilon = 1$, the method is equivalent to the graph coloring method; if $\varepsilon = 2$, the method becomes a variation of color mixing methods; if ε equals the maximum extent found in G, the method becomes similar to the structural method with behavioral STVs. A graph enhancing algorithm of complexity $O(N^3)$ has been described and implemented. It is also explained how the complexity of the algorithm can be improved to $O(D^{\varepsilon} \times N)$.

7.5 Diagnostic resolution assessment

A methodology for quantitative assessment of diagnostic resolution (DR) has been proposed. The new method can grade any diagnostic scheme, behavioral or structural, before the scheme is used in a real process. The assessment technique is based on evaluating the diagnosability of each individual fault in a fault sample, and then estimating DR as the weighted average fault diagnosability of the sample. Each individual fault is weighted with its occurrence probability, which depends on its multiplicity. The fault multiplicity distribution is a negative binomial distribution whose parameters are obtained from a statistical characterization of the process.

The DR figure of merit intends to be a measure of fault isolation efficiency. The individual fault diagnosability tries to reflect the ratio between the fault isolation time with a perfect diagnosis by the fault isolation time with the actual diagnosis. It is assumed that this ratio is roughly given by the ratio of the number of single faults that have actually occurred by the number of single suspect faults in the diagnosis.

The principal difficulty of the DR assessment method is the criterion for fault sampling. The quality of the DR figure of merit is obviously sensitive to the faults selected to be part of the sample. Our fault sampling criterion is based on the analysis of empirical facts observed in several manufacturing processes. After a careful examination of these facts, three main assumptions have been made: (1) the occurrence of short and open faults are independent processes; (2) open faults do not cause diagnostic ambiguity; (3) shorts that form connected subgraphs are more likely than other short topologies. Based on these assumptions the following fault sampling criterion has been formulated: the fault sample should consist of only shorts that form connected subgraphs and whose multiplicity does not exceed a certain value r_s^{max}. Although open faults are not included in the sample, it is assumed they have diagnosability 1, and the contribution (γ_o) they give to DR is accounted for. The shorts diagnosability DR_s is defined as the weighted average short diagnosability, and is estimated from the sample. The final DR value is a function of γ_o and DR_s.

The methodology for DR assessment is implemented by an interconnect diagnosis simulator. The simulator inputs the adjacency graph G, the test set

MTV and the statistical parameters that characterize the faults. Then it computes the *DR* figure of merit and outputs several other data such as the process yield Y, the maximum short multiplicity r_s^{max}, etc. The complexity of the simulator algorithm is $O(r_s^{max} \times N_{r_s} \times N_b \times p \times N)$ for a behavioral scheme and $O(r_s^{max} \times N_{r_s} \times N_b \times p \times r_s^{max} \times D)$ for a structural scheme.

7.6 Experimental results

The interconnect diagnosis methods have been applied to six PCB examples and results on fault extraction, diagnostic synthesis and diagnosis simulation have been obtained. This study has been carried out using a statistical process characterization obtained from the results published for an actual SMT process (Tegethoff, 1994a). The results contemplate the traditional behavioral methods and the recent structural methods. The three types of structural methods are analyzed: with behavioral, graph coloring and statistical color mixing STVs. For each method three representative STV sequences have been used: the modified counting counting sequence, the self-diagnosis and the N sequence.

The fault extraction algorithm has been shown to perform efficiently, taking only a few minutes for circuits of a moderate size (thousands of nets). For very large EAs (tens of thousands of nets), the program may take a few days. This may be acceptable since fault extraction is only performed once. If more efficiency is required, the improvements suggested for the fault extraction algorithm can significantly boost its performance; it is estimated that with the improved version the largest EAs would run in a few seconds.

The traditional behavioral methods have been shown to produce significantly longer STVs than the structural methods. The statistical color mixing method can produce both shorter STVs and superior diagnostic resolution. These results show that there is much room for optimizing the methods currently used in the industry. One simple improvement is a method that combines behavioral STVs with structural diagnostic analysis. This method produces the best possible diagnostic resolution. In fact, structural diagnostic analysis resolves many situations which would be ambiguous in behavioral diagnosis, and prevents fault diagnosability from degrading rapidly with fault multiplicity.

Structural diagnosis with graph coloring STVs can produce the sortest STVs while still offering minimal diagnostic capabilities for circuits of a medium size or lower. However, as the circuit size increases, higher fault multiplicities become more likely, causing the individual fault diagnosabilities to decay faster for the behavioral method than for the graph coloring method. This makes the graph coloring method to become better than the behavioral method for large circuit sizes.

Structural diagnosis with statistical color mixing STVs can optimize both the STV length and the diagnostic resolution. Compared to the graph coloring method, the statistical color mixing STVs are slightly longer. However, in terms of *DR*, the method is considerably better than the graph coloring method. Its *DR* is only comparable to the diagnostic resolution of the structural method with behavioral STVs, i.e., it is close to the maximum achievable *DR*.

Experimental results have demonstrated that the computational cost associated with structural interconnect diagnosis is very affordable. The computer programs implemented for fault extraction, graph coloring, graph enhancing and diagnosis simulation have efficiently run examples of moderate sizes on a low cost computer. It has been shown that the improved versions of algorithms can run quite large EAs in a matter of seconds or, in the worst case, in a few minutes.

7.7 Future work

In this research we have thoroughly analyzed the problem of testing and diagnosing interconnect faults in boundary-scan electronic assemblies. Interconnect circuits are the simplest circuits that can be imagined; at least for these circuits we hope to have given satisfactory answers to many of the problems that arise in test and diagnosis. The challenge now is in applying the same rigor and discipline to more complex circuits and systems. As far as interconnect faults are concerned, a few problems have been left open, which we will discuss in the following paragraphs.

Some assumptions about the fault models used need to be verified. Namely, more experimental studies should be conducted in order to verify whether the fault models adequately represent the actual faults. Do other fault behaviors need to be considered? What is the probability to be assigned to each fault behavior? Fault modeling is the backbone of any successful testing and diagnosis approach. Thus, it is essential to make sure that we are working with the right models.

The study of behavioral test vector sequences is interesting from a theoretical point of view. These sequences are used both in behavioral and structural diagnostic synthesis. In the former, the sequences are applied to the nets; in the latter, they are applied to the colors. In particular, the aspect of optimality has been left open; it is still not known what are the optimal sequences for avoiding diagnostic ambiguity.

This work has demonstrated that considering realistic faults and weighting them with their occurrence probabilities can lead to better test and diagnosis schemes, which do not necessarily have a prohibitive cost. Due to the simplicity of interconnect circuits, we have been able to construct a solid test and diagnosis methodology. This study provides a good starting point when considering test and diagnosis methodologies for more complex systems. In the end, the problems encountered are always the same: which fault models should be used? How can test and diagnosis vectors be generated? Can the amount of test data be reduced? Can lower level information such as layout information be used efficiently? How do fault coverage and diagnostic resolution relate to product quality? What is the cost of testing and diagnosing? More importantly, what is the cost of not testing and not diagnosing?

Appendix A
Layout file format

In this appendix the layout file format used by the Pcb software package is described, and an example file is given.

A.1 Format description

```
File            = Header Font PCBData
Header          = PCBName [GridData] [CursorData] [PCBFlags] [Groups]
PCBName         = "PCB(" Name Width Height ")"
GridData        = "Grid(" Grid GridOffsetX GridOffsetY ")"
CursorData      = "Cursor(" X Y zoom ")"
PCBFlags        = "Flags(" Flags ")"
Groups          = "Groups(" GroupString ")"
Font            = @{FontData@}...
FontData        = @{Symbol@}...
Symbol          = "Symbol(" SymbolID Spacing ")"
                      "(" @{SymbolData@}... ")"
SymbolData      = @{SymbolLine@}...
SymbolLine      = "SymbolLine(" X1 Y1 X2 Y2 Thickness ")"
PCBData         = @{Via | Layer | Element@}...
Via             = "Via(" X Y Thickness DrillingHole Name Flags ")"
Element         = "Element(" Flags CanonicalName LayoutName \
                      TextX TextY direction scale TextFlags")"
                      "(" @{ElementData@}... [Mark] ")"
ElementData     = @{ElementLine | Pin | ElementArc @}...
ElementLine     = "ElementLine(" X1 Y1 X2 Y2 Thickness ")"
Pin             = "Pin(" X Y Thickness DrillingHole Name Flags ")"
ElementArc      = "ElementArc(" X Y Width Height
                      StartAngle DeltaAngle Thickness ")"
Mark            = "Mark(" X Y ")"
Layer           = "Layer(" LayerNumber Name ")"
                      "(" @{LayerData@}... ")"
```

```
LayerData       = @{Line | Polygon | Text@}...
Line            = "Line(" X1 Y1 X2 Y2 Thickness Flags")"
Polygon         = "Polygon(" Flags ")" \
                       "(" @{Points@}... ")"
Points          = "(" X Y ")"
Text            = "Text(" X Y direction scale TextData Flags")"
```

A.2 File example

```
PCB("ir232" 20000 20000)

Grid(100 80 55)
Cursor(30 590 2)
Flags(0x0000)
Groups("1:2:3:4:5:6:7:8")

Symbol(' ' 20)
(
        SymbolLine(0 53 40 53 1)
)
```

(stuff deleted...)

```
Element(0x0000 "LED" "LED3" 1730 2060 0 100 0x0000)
(
        Pin(1790 1940 80 32 "A" 0x0001)
        Pin(1890 1940 80 32 "K" 0x0001)
        ElementLine (1940 1890 1940 1990 15)
        ElementArc (1840 1940 110 110 210 300 15)
        Mark (1890 1940)
)
```

(stuff deleted...)

```
Layer(1 "trace")
(
Line(960 930 1010 930 20 0x0000)
Line(830 800 960 930 20 0x0000)
```

(stuff deleted...)

```
Text(1720 1080 0 100 "+" 0x0000)
Text(430 2090 1 100 "IR232/JOHN DUBOIS/95-01" 0x0000)
)
```

Appendix B
Graph coloring

In this appendix, the graph coloring algorithm used in the diagnostic synthesis algorithms of Section 4.4 is described. The algorithm inputs a graph $G = (C, E)$ and computes a color mapping $CM(C) : C \rightarrow CS$, where CS is a set of colors. For a net n_i, its color is denoted as $CM(n_i)$. A greedy heuristic to minimize the total number of colors k is used. This algorithm was originally proposed by Matula et al., and shown by Garey et al. to be appropriate for coloring adjacency graphs derived from PCB layouts (Matula et al., 1972; Garey et al., 1976). The algorithm uses a recursive procedure outlined below:

Algorithm B.1 (colorGraph(G)) :

1. n_i = *selectMinimumDegreeNet(G);*
2. G' = *removeVertexFromGraph(*n_i*, G);*
3. **if** *(G! =* $\emptyset$*)* {
4. *colorGraph(G');*
5. $CM(n_i)$ = *lowestColorUnusedByNeighbors(G);* }
6. **return** CM*;*

The algorithm goes through N recursion levels, that is, one level for each removed net. At each recursion level the most complex step is the step of selecting the net with the minimum number of neighbors or degree (line 1). The algorithm is implemented using the incidence matrix representation $A(G)$ (see Section 4.1.3). Using $A(G)$, the complexity of the step in line 1 is $O(N^2)$, since it consists of counting the 1s in each row of $A(G)$. Because line 1 is executed for all N recursion levels, the overall complexity of Algorithm B.1 is $O(N^3)$.

It is possible to improve the performance of the algorithm by initially sorting the nets in graph G by their degree, and re-sorting after each net removal (step 2). To do the sortings, graph G is best represented using a linked list data structure $L(G)$, also mentioned in Section 4.1.3. If the nets are sorted, selecting the net with the minimum degree becomes trivial: just pick the head or tail of $L(G)$, depending on whether the list is sorted in ascending or descending order, respectively. The complexity of the initial sorting is $O(D \times N)$ using, for example, an $N \times D_{max}$ table. The complexity of re-sorting is simply $O(D)$ because only the neighbors of the last net removed need reinserting. Therefore, the overall complexity of the algorithm becomes $O(D \times N)$.

Garey et al. also showed that, with Algorithm B.1, a maximum of $D_{max} + 1$ colors can be obtained, that is, $k_{min} \leq D_{max} + 1$. This shows that $k \ll N$, since, usually, $D_{max} \ll N$. It should be noted that this is a conservative upper bound; there are theoretically less conservative upper bounds (McHugh, 1990), which are closer to the result normally obtained in practice.

Appendix C
Graph enhancing

In this appendix two graph enhancing algorithms are presented. The graph enhancing procedure adds extra edges to graph G, in order to produce an enhanced graph G_ε. An extra edge is added between any two vertices whose edge distance is ε or lower. Algorithm 4.4 and Algorithm 4.5, studied in Sections 4.4.3 and 4.4.4, respectively, use graph enhancing before graph coloring to improve the diagnostic capability of test sets.

The first graph enhancing algorithm to be discussed is the one used in this work to obtain the experimental results of Section 6. Its input is the incidence matrix $A(G)$, and output is the incidence matrix of graph G_ε.

Algorithm C.1 (enhanceGraph $(G,\ \varepsilon)$**)** :
1. $A(G_\varepsilon) = \sum_{i=1}^{\varepsilon}[A(G)]^i$;
2. *normalize*$(A(G_\varepsilon))$;
3. **return** $A(G_\varepsilon)$;

It can be shown that after the step in line 1, any two vertices n_i and n_j will have produced a non-zero element in the incidence matrix $A_{i,j}(G_\varepsilon)$ of the graph G_ε. The non-zero element means that edge $e_{i,j}$ exists in G_ε. To implement the step in line 1, one needs $\varepsilon - 1$ matrix sums ($O(N^2)$) and $\varepsilon - 1$ matrix multiplications ($O(N^3)$). The normalize step in line 2 simply puts G_ε in the usual incidence matrix form, where any non-zero element is normalized to 1. The complexity of this step is $O(N^2)$, since all N^2 elements of $A(G_\varepsilon)$ are checked. Therefore, the step in line 1 determines the overall $O(\varepsilon \times N^3)$ time complexity of Algorithm C.1. To help understand how Algorithm C.1 works, the following example is presented:

Example C.1 *In this example, the graphs $A(G_2)$ and $A(G_3)$ for the graph G in Fig. 4.2(b) are computed. To begin with, the incidence matrix $A(G)$ of G is* *written as:*

$$A = \begin{bmatrix} 0 & 1 & 0 & 0 & 0 \\ 1 & 0 & 1 & 1 & 0 \\ 0 & 1 & 0 & 1 & 1 \\ 0 & 1 & 1 & 0 & 0 \\ 0 & 0 & 1 & 0 & 0 \end{bmatrix}$$

Then the matrices $A(G)$ to the powers 2 and 3, that is, $[A(G)]^2$ and $[A(G)]^3$, are computed. The result is given below:

$$[A(G)]^2 = \begin{bmatrix} 1 & 0 & 1 & 1 & 0 \\ 0 & 3 & 1 & 1 & 1 \\ 1 & 1 & 3 & 1 & 0 \\ 1 & 1 & 1 & 2 & 1 \\ 0 & 1 & 0 & 1 & 1 \end{bmatrix} \quad [A(G)]^3 = \begin{bmatrix} 0 & 3 & 1 & 1 & 1 \\ 3 & 2 & 5 & 4 & 1 \\ 1 & 5 & 2 & 4 & 3 \\ 1 & 4 & 4 & 2 & 1 \\ 1 & 1 & 3 & 1 & 0 \end{bmatrix}$$

Finally, the matrices $[A(G)]^2$ and $[A(G)]^3$ are normalized to obtain matrices $A(G_2)$ and $A(G_3)$:

$$A(G_2) = \begin{bmatrix} 0 & 1 & 1 & 1 & 0 \\ 1 & 0 & 1 & 1 & 1 \\ 1 & 1 & 0 & 1 & 1 \\ 1 & 1 & 1 & 0 & 1 \\ 0 & 1 & 1 & 1 & 0 \end{bmatrix} \quad A(G_3) = \begin{bmatrix} 0 & 1 & 1 & 1 & 1 \\ 1 & 0 & 1 & 1 & 1 \\ 1 & 1 & 0 & 1 & 1 \\ 1 & 1 & 1 & 0 & 1 \\ 1 & 1 & 1 & 1 & 0 \end{bmatrix}$$

To see that the result is correct note that $A(G_2)$ is the incidence matrix of graph G_2 in Fig. 4.6, and that matrix $A(G_3)$ represents a fully connected graph, since the maximum edge-distance in G is $ED(n_1, n_5) = 3$.

If the linked list representation $L(G)$ is used, it is possible to speed up the search for nets at edge-distance ε or lower. The second algorithm, Algorithm C.2, improves upon Algorithm C.1 by doing that and can be described as follows:

Algorithm C.2 (enhanceGraph (G, ε)) :
1. $G_\varepsilon = G$;   `/* initialize` G_ε `*/`
2. **for each** `/* net */` $n_i \in G$ **do** {
3. $\quad I$ = *getNetsWithinMaxStainExtent*(n_i, ε, G);
4. $\quad$ **for each** `/* net */` $n_j \in I$ **do**
5. $\qquad G_\varepsilon$ = *insertEdgeInGraph*(G_ε, $e_{i,j}$);
6. **return** G_ε;

The complexity of Algorithm C.2 is bounded by the complexity of the function call *getNetsWithinMaxStainExtent*(n_i, $e_{i,j}$, G), which returns the set of nets I within edge-distance ε from net n_i. That is, $\forall_{n_j \in I} : ED(n_i, n_j) \leq \varepsilon$. Function *getNetsWithinMaxStainExtent()* can be implemented using a recursive procedure such as the one outlined below:

Algorithm C.3 (getNetsWithinMaxStainExtent(n_i, ε, G)) :
1. $I = \emptyset$;   `/* initialize` I `*/`

```
for each /* net */ n_j ∈ neighborhoodOf(n_i) do
    if (ε > 1)
        I = I ∪ getNetsWithinMaxStainExtent(n_i, ε − 1, G);
    I = I ∪ n_j;
return I;
```

The complexity of Algorithm C.3 is $O(D'^{\varepsilon})$, where D' is approximately the same as D, the average vertex degree of G. The difference between D and D' is that the former is an arithmetic average, whereas the latter is a geometric one. Note that the recursive procedure accounts for the neighbors of neighbors, etc., down to depth ε. Thus, the vertex degrees get multiplied by themselves ε times, and hence D' should in rigor be the geometric average. Nevertheless, if ε is small, then D and D' are practically equal.

Since in Algorithm C.2 function *getNetsWithinMaxStainExtent()* is called N times with arguments n_i, ε, G, therefore the overall complexity of Algorithm C.2 is $O(N \times D^{\varepsilon})$. The performance Algorithm C.2 is substantially better than the performance of Algorithm C.1, for in practice ε is small, and $N \times D^{\varepsilon} < N^3$.

References

Abramovici, M., Breuer, M. A., and Friedman, A. D. (1990). *Digital Systems Testing and Testable Design.* Computer Science Press.

Agrawal, V. D., Seth, S. C., and Agrawal, P. (1982). "Fault Coverage Requirement in Production Testing of LSI Circuits". *IEEE Journal of Solid State Circuits*, SC-17(1):57–61.

Aitken, R. C. (1995). "Finding Defects with Fault Models". In *Proc. Int. Test Conference (ITC)*, pages 498–505.

Appel, K. and Haken, W. (1977). "The Solution of the Four-Colour Problem". *Scientific American*, 27(4):108–121.

Barnfield, S. J. and Moore, W. R. (1989). "Multiple Fault Diagnosis in Printed Circuit Boards". In *Proc. Int. Test Conference (ITC)*, pages 662–671.

Basset, R. W., Butkus, B. J., Dingle, S. L., Faucher, M. R., Gillis, P. S., Panner, J. H., J. G. Petrovick, J., and Wheather, D. L. (1990). "Low-Cost Testing of High-Density Logic Components". *IEEE Design and Test of Comp.*, 7(2):15–28.

Bateson, J. (1985). *In-Circuit Testing.* Van Nostrand Reinhold Company.

Bleeker, H., van den Eijnden, P., and de Jong, F. (1993). *Boundary-Scan Test: A Practical Approach.* Kluwer Academic Publishers.

Bose, B. and Rao, T. R. N. (1982). "Theory of Unidirectional Error Correcting/Detecting Codes". *IEEE Trans. Comp.*, C-31:521–530.

Buckroyd, A. (1994). *In-Circuit Testing.* Butterworth-Heinemann.

Chen, C. C. and Hwang, F. K. (1989). "Detecting and Locating Electrical Shorts Using Group Testing". *IEEE Trans. on Circuits and Systems*, CAS-36(8):1113–1116.

Chen, X. T. and Lombardi, F. (1996). "A Coloring Approach to the Structural Diagnosis of Interconnects". In *Proc. Int. Conf. on Comp. Aided Design (ICCAD)*, pages 676–680.

Chen, X. T. and Lombardi, F. (1998). "Structural Diagnosis of Interconnects by Colouring". *ACM Trans. on Design Automation of Electronic Systems*, 3(2):249–271.

Cheng, W.-T., Lewandowski, J. L., and Wu, E. (1990). "Diagnosis for Wiring Interconnects". In *Proc. Int. Test Conference (ITC)*, pages 565–571.

Crook, D. T. (1979). "Analog In-Circuit Component Measurements: Problems and Solutions". *Hewlett-Packard Journal*, pages 19–22.

Crook, D. T. (1990). "A Fourth Generation Analog In-Circuit Program Generator". In *Proc. Int. Test Conference (ITC)*, pages 605–612.

Cunningham, J. A. (1990). "The Use and Evaluation of Yield Models in Integrated Circuits Manufacturing". *IEEE Trans. on Semiconductor Manufacturing*, 3(2):60–71.

de Gyvez, J. P. (1991). *IC Defect-Sensitivity*. PhD thesis, Tech. Univ. Eindhoven.

de Jong, F. (1991). "Testing the Integrity of the Boundary-Scan Test Infrastructure". In *Proc. Int. Test Conference (ITC)*, pages 106–112.

de Jong, F. and van Wijngaarden, A. (1992). "Memory Interconnection Test at Board Level". In *Proc. Int. Test Conference (ITC)*, pages 328–337.

de Sousa, J. T. (1996). "Analysis, Test and Diagnosis of Interconnect Faults in Printed Circuit Boards". MPhil to PhD transfer report, Imperial College of Science, Technology and Medicine.

de Sousa, J. T. (1998). *Diagnosis of Interconnect Defects in Electronic Assemblies*. PhD thesis, Imperial College of Science, Technology and Medicine.

de Sousa, J. T. and Agrawal, V. D. (2000). Reducing the complexity of defect level modeling using the clustering effect. In *Proc. Design, Automation and Test in Europe (DATE) Conf.*, pages 640–644.

de Sousa, J. T. and Cheung, P. Y. K. (1997a). "Diagnosis of Boards for Realistic Interconnect Shorts". *Journal of Electronic Testing: Theory and Applications (JETTA)*, 11(2):157–171.

de Sousa, J. T. and Cheung, P. Y. K. (1997b). "Diagnosis of Realistic Interconnect Shorts". In *Proc. European Design and Test Conference (ED&TC)*, pages 501–506.

Feng, C., Huang, W., and Lombardi, F. (1995). "A New Diagnosis Approach for Short Faults in Interconnects". In *IEEE Fault Tolerant Computing Symposium (FTCS)*, pages 331–339.

Ferguson, F. J. and Shen, J. P. (1988). "A CMOS Fault Extractor for Inductive Fault Analysis". *IEEE Trans. on Computer Aided Design*, 7(11):1181–1194.

Ferris-Prabhu, A. V. (1985). "Modeling the Critical Area in Yield Forecast". *IEEE J. Solid-State Circuits*, SC-20(4):874–877.

Freiman, C. V. (1962). "Optimal Error Detection Codes for Completely Asymmetric Binary Channels". *Information and Control*, 5:64–71.

Galiay, J., Crouzet, Y., and Vergniault, M. (1980). "Physical Versus Logical Fault Models in MOS LSI Circuits: Impact on Their Testability". *IEEE Trans. on Comp.*, C-29(6):527–531.

Gandemer, S., Tremintin, B. C., and Charlot, J.-J. (1988). "Critical Area and Critical Levels Calculation in I.C. Yield Modeling". *IEEE Trans. Electron Devices*, 35(2):158–166.

Garey, M. R. and Johnson, D. S. (1979). *Computers and Intractability: A Guide to Theory of NP-Completeness.* W. H. Freeman & Co.

Garey, M. R., Stifler, D. S., and So, H. C. (1976). "An Application of Graph Coloring to Printed Circuit Testing". *IEEE Trans. on Circuits and Systems*, CAS-23(10):591–599.

Goel, P. and McMahon, M. T. (1982). "Electronic Chip in Place Test". In *Proc. Int. Test Conference (ITC)*, pages 83–90.

Hassan, A., Agarwal, V. K., Rajski, J., and Dostie, B. N. (1989). "Testing of Glue Logic Interconnects using Boundary Scan Architecture". In *Proc. Int. Test Conference (ITC)*, pages 700–711.

Hassan, A., Rajski, J., and Agarwal, V. K. (1988). "Testing and Diagnosis of Interconnects using Boundary Scan Architecture". In *Proc. Int. Test Conference (ITC)*, pages 126–137.

Hedelsbrunner, H. (1987). *Algorithms in Combinatorial Geometry.* Springer-Verlag.

Her, W. C., Jin, L. M., and El-Ziq, Y. (1992). "An ATPG Driver Selection Algorithm for Interconnect Test with Boundary Scan". In *Proc. Int. Test Conference (ITC)*, pages 382–388.

Huisman, L. (1995). "Yield Fluctuations and Defect Models". *J. of Electron. Testing: Theory and Applications (JETTA)*, 7(3):241–254.

IEEE (1990). "1149.1 Standard Access Port and Boundary-Scan Architecture".

IEEE (1995). "1149.5 Standard Module Test and Maintenance Bus".

IEEE (1999). "149.4 Standard Mixed-Signal Test Bus".

Jarwala, N. and Yau, C. W. (1989). "A New Framework for Analyzing Test Generation and Diagnosis Algorithms for Wiring Interconnects". In *Proc. Int. Test Conference (ITC)*, pages 63–70.

Jarwala, N. and Yau, C. W. (1991). "Achieving Board-Level BIST Using the Boundary-Scan Master". In *Proc. Int. Test Conference (ITC)*, pages 649–658.

Jarwala, N., Yau, C. W., Stiling, P., and Tammaru, E. (1992). "A Framework for Boundary-Scan Based System Test and Diagnosis". In *Proc. Int. Test Conference (ITC)*, pages 993–998.

Johnson, B. (1993). "Boundary Scan Eases Test of New Technologies". *Test & Measurement Europe*, pages 25–30.

Jones, O. (1989). *Introduction to the X Window System.* Prentice-Hall.

Kautz, W. H. (1974). "Testing for Faults in Wiring Networks". *IEEE Transactions on Computers*, C-23(4):358–353.

Kernighan, B. W. and Ritchie, D. M. (1978). *The C Programming Language.* Prentice-Hall.

Landis, D., Hudson, C., and McHugh, P. (1992). "Applications of the IEEE P1149.5 Module Test and Maintenace Bus". In *Proc. Int. Test Conference (ITC)*, pages 984–992.

Lee, N.-C. (1993). "Pratical Considerations for Mixed-Signal Test Bus". In *Proc. Int. Test Conference (ITC)*, pages 591–592.

Leung, Y.-W. (1993). "A Signal Path Grouping Algorithm for Fast Detection of Short Circuits on Printed Circuit Boards". *IEEE Trans. on Inst. and Measur.*, 43(1):80–85.

Lewandowski, J. L. and Velasco, V. J. (1986). "Short Circuit Testing". Technical Report CC8559, AT&T.

Lien, J. C. and Breuer, M. A. (1991). "Maximal Diagnosis for Wiring Networks". In *Proc. Int. Test Conference (ITC)*, pages 96–105.

Liu, T., Chen, X. T., Lombardi, F., and Salinas, J. (1996). "Layout-driven Detection of Bridge Faults in Interconnects". In *Proc. IEEE Workshop on Defect and Fault Tolerance in VLSI Systems*, pages 105–113.

Lubaszewski, M. and Courtois, B. (1992). "On The Design of Self-Checking Boundary Scannable Boards". In *Proc. Int. Test Conference (ITC)*, pages 372–381.

Matos, J. S., Leao, A. C., and Ferreira, J. C. (1993). "Control and Observation of Analog Nodes in Mixed-Signal Boards". In *Proc. Int. Test Conference (ITC)*, pages 323–331.

Matula, D. M., Marble, G., and Isaacson, J. D. (1972). *Graph Coloring Algorithms, in Graph Theory and Computing.* Academic Press.

Maunder, C. and Beenker, F. (1987). "Boundary Scan: A framework for structured design-for-test". In *Proc. Int. Test Conference (ITC)*, pages 714–723.

Maunder, C. M. and Tulloss, R. E., editors (1990). *The Test Access Port and Boundary-Scan Architecture.* IEEE Computer Society Press.

McBean, D. and Moore, W. R. (1993). "Testing Interconnects: A Pin Adjacency Approach". In *Proc. European Test Conference (ETC)*, pages 484–490.

McHugh, J. A. (1990). *Algorithmic Graph Theory.* Prentice-Hall International Editions.

Melton, M. and Brglez, F. (1992). "Automatic Pattern Generation for Diagnosis of Wiring Interconnect Faults". In *Proc. Int. Test Conference (ITC)*, pages 389–398.

Michel, W. (1994). "PCI-Implementation of a SCSI Computer Peripheral Controller". *Elektronik*, 43(26):81–84.

Nau, T. (1995a). *"PCB: an interactive PCB layout system for X11 - Source code in C and X11"*. Software Version 1.3.

Nau, T. (1995b). *"PCB: an interactive PCB layout system for X11 - User's Manual"*. Software Version 1.3.

Osseiran, A., editor (1999). *Analog and Mixed-Signal Boundary-Scan A Guide to the IEEE 1149.4 Test Standard.* Kluwer Academic Publishers, Boston.

Park, S. (1996). "A New Complete Diagnosis Patterns for Wiring Interconnects". In *33rd ACM/IEEE Design Autom. Conf. (DAC)*.

Parker, K. P. (1992). *The Boundary-Scan Handbook.* Kluwer Academic Publishers.

Parker, K. P. (1998). *The Boundary-Scan Handbook.* Kluwer Academic Publishers, Boston, second edition.

Parker, K. P., McDermit, J. E., and Oresjo, S. (1993). "Structure and Metrology for an Analog Testability Bus". In *Proc. Int. Test Conference (ITC)*, pages 309–322.

Rogel-Favila, B. (1991). *Model-Based Fault Diagnosis of Digital Circuits.* PhD thesis, Imperial College of Science, Technology and Medicine.

Rogel-Favila, B. (1992). "Automatic Test Generation and Fault Diagnosis of Boundary Scan Circuits". In *IEE Colloquium on Automated Testing and Software Solutions*, pages 3/1–4.

Rogers, A. (1974). *Statistical Analysis of Spatial Dispersion.* London, UK: Pion, Ltd.

Salinas, J., Shen, Y., and Lombardi, F. (1996). "A Sweeping Line Approach to Interconnect Testing". *IEEE Trans. on Comp.*, C-45(8):917–929.

Sedgewick, R. (1983). *Algorithms.* Addison-Wesley Publishing Company.

Seth, S. C. and Agrawal, V. D. (1984). "Characterizing the LSI Yield Equation from Wafer Test Data". *IEEE Trans. on CAD*, CAD-3(2):123–126.

Shen, J. P., Maly, W., and Ferguson, F. J. (1985). "Inductive Fault Analysis of NMOS and CMOS Circuits". *IEEE Design and Test of Computers*, 2:13–26.

Shi, W. and Fuchs, W. K. (1995). "Optimal Interconnect Diagnosis of Wiring Networks". *IEEE Trans. on VLSI Systems*, 3(3):430–436.

Sousa, [1] J. J. T., Goncalves, F. M., and Teixeira, J. P. (1991). "IC Defects-Based Testability Analysis". In *Proc. Int. Test Conf. (ITC)*, pages 500–509.

Sousa, J. J. T., Goncalves, F. M., Teixeira, J. P., Marzocca, C., Corsi, F., and Williams, T. W. (1996a). "Defect Level Evaluation in an IC Design Environment". *IEEE Trans. on CAD*, 15(10):1286–1293.

Sousa, J. T., Shen, T., and Cheung, P. Y. K. (1996b). "On Structural Diagnosis for Interconnects". In *Proc. Int. Symp. on Circuits and Systems*, pages 532–535.

Sousa, J. T., Shen, T., and Cheung, P. Y. K. (1996c). "Realistic Fault Extraction for Boards". In *Proc. European Design and Test Conference (ED&TC)*, page 612.

Stapper, C. H., Armstrong, F., and Saji, K. (1983). "Integrated Circuit Yield Statistics". *Proc. IEEE*, 71:453–470.

Su, C. and Jou, S.-J. (1999). "Decentralized BIST Methodology for System Level Interconnects". *Journal of Electronic Testing: Theory and Applications (JETTA)*, 15(3):255–265.

Tegethoff, M. M. V. (1994a). "Defects, Fault Coverage, Yield and Cost in Board Manufacturing". In *Proc. Int. Test Conference (ITC)*, pages 539–547.

Tegethoff, M. M. V. (1994b). "Manufacturing Test Simulator: A Concurrent Engineering Tool for Boards and MCMs". In *Proc. Int. Test Conference (ITC)*, pages 903–910.

Teixeira, J. P., Goncalves, F. M., and Sousa, J. J. T. (1990). "Realistic Fault List Generation for Physical Testability Assessment". In *Proc. IEEE Workshop on Defect and Fault Tolerance in VLSI Systems*, pages 131–140.

[1] Some papers of J. T. de Sousa have been published under the name J. (J.) T. Sousa

Thatcher, C. W. and Tulloss, R. E. (1993). "Towards a Test Standard for Board and System Level Mixed-Signal Interconnects". In *Proc. Int. Test Conference (ITC)*, pages 300–308.

Vaucher, C. and Balme, L. (1993). "The Standard Mirror Boards (SMBs) Concept". In *Proc. Int. Test Conference (ITC)*, pages 672–679.

Vinnakota, B., editor (1998). *Analog and mixed-signal test.* Prentice-Hall PTR.

Wadsack, R. L. (1978). "Fault Modeling and Logic Simulation of CMOS and NMOS Integrated Circuits". *Bell Syst. Tech. Journal*, 57(2):1449–74.

Wagner, P. T. (1987). "Interconect Testing with Boundary Scan". In *Proc. Int. Test Conference (ITC)*, pages 52–57.

Wassink, R. J. K. (1989). *Soldering in Electronics.* Electrochemical Publications Limited.

Williams, T. W. (1982). "Design For Testability - A Survey". *IEEE Transactions On Computers*, C-31(1):1–15.

Williams, T. W. and Brown, N. C. (1981). "Defect Level as a Function of Fault Coverage". *IEEE Transactions on Computers*, C-30(12):987–988.

Yau, C. W. and Jarwala, N. (1989). "A Unified Theory for Designing Optimal Test Generation and Diagnosis Algorithms for Board Interconnects". In *Proc. Int. Test Conference (ITC)*, pages 71–77.

Young, D. A. (1990). *The X Window System: Programming and Applications with Xt.* Englewood Cliffs, New Jersey: Prentice-Hall.

Abbreviations and acronyms

AB1: on-chip analog test bus number 1.

AB2: on-chip analog test bus number 2.

ASIC: application specific circuit.

AT1: off-chip analog test bus number 1.

AT2: off-chip analog test bus number 2.

ATPG: automatic test pattern generation.

A-R: any response fault model.

ATE: automatic test equipment.

BIST: built-in self-test.

BS: boundary-scan.

BSEA: boundary-scan electronic assembly.

BSC: boundary-scan cell.

BSR: boundary-scan register.

CATE: cost of automatic test equipment.

CAR: computer-aided repair.

CD: cost of diagnosis.

CDI: cost of diagnosis and fault isolation.

CMOS: complementary metal-oxide-semiconductor.

COB: chip on board.

CRR: cost of repair and retest.

CPLD: complex programmable logic device.

CTA: cost of test application.

CUT: circuit under test.

DFT: design for testability.

DR: diagnostic resolution.

DL: defect level.

EA: electronic assembly.

ECL: emitter coupled logic.

EDA: electronic design automation.

EXTEST: boundary-scan instruction for testing the external circuitry.

FIE: fault isolation efficiency.

FPGA: field programmable gate array.

I/O: input/output.

IC: integrated circuit.

ICT: in-circuit testing.

IEEE: institute of electrical and electronic engineers.

IFA: inductive fault analysis.

INTEST: boundary-scan instruction for testing the internal logic core of a chip.

IP: intellectual property.

MCM: multi-chip module.

MRV: matrix of response vectors.

MTV: matrix of test vectors.

PC: personal computer.

PCI: peripheral component interconnect.

PCB: printed circuit board.

PhD: doctor of philosophy.

POST: power-on self-test.

PS: pin spacing.

PSN: primary shorting nets.

PTV: parallel test vector.

PV: production volume.

RUNBIST: boundary-scan instruction for running chip self-tests.

S-A-0/1: stuck at logic 0/1 fault model.

S-D-k: strong driver net k (n_k) fault model.

S-R: same response fault model.

SFR: solder fault rate.

SOC: system on a chip.

SOR: solder open rate.

SSR: solder short rate.

SMT: surface mount technology.

SSN: secondary shorting nets.

STV: serial test vector.

TAB: tape automated bonding.

TAP: boundary-scan test access port.

TCK: boundary-scan test clock.

T&D: test and diagnosis.

TBIC: test bus interface circuit.

TDI: boundary-scan test data input.

TDO: boundary-scan test data output.

TMS: boundary-scan test mode select.

TTL: transistor-transistor logic.

VDT: video display terminal.

W-A/O: wired-and/or fault model.

WITD: wire interconnect test and diagnosis.

Notation

$\underline{0}$: Boolean vector given by a sequence of 0s.

$\underline{1}$: Boolean vector given by a sequence of 1s.

$A(G)$: incidence matrix representation of G.

C: set of all nets in a generic interconnect circuit.

$CATE$: cost of automatic test equipment.

CD: cost of diagnosis.

CDI: cost of diagnosis and fault isolation.

CRR: cost of repair and retest.

CS: color set.

CTA: cost of test application.

CU: unitary cost of test, diagnosis and repair.

D: average vertex degree of graph G.

DR: diagnostic resolution.

DR': estimate of diagnostic resolution.

DR_s: shorts' diagnostic resolution.

DR'_s: estimate of shorts' diagnostic resolution.

D_{max}: maximum vertex degree of graph G.

E: the set of all single short faults considered.

E': set of edges in a stain.

E'': set of edges in a faulty subgraph.

$E(n_i, n_j)$: edge-distance between nets n_i and n_j.

$EXT(T)$: extent of graph T.

F: a generic fault, i.e., a non-empty set of single faults.

F': diagnosis of fault F.

$F_j^{r_o}$: a particular multiplicity r_o open fault.

$F_{i,j}^{r_s,r_o}$: multiple mixed fault formed by multiple short $F_i^{r_s}$ and multiple open $F_j^{r_o}$.

$F_i^{r_s}$: a particular multiplicity r_s short fault.

FIE: fault isolation efficiency.

FIE_F: individual fault isolation efficiency for fault F.

G: adjacency graph ($G = (C, E)$).

G_ε: graph obtained from graph G with Algorithm 4.5.

H: faulty subgraph ($H = (V, E'')$).

$H(STV)$: Hamming weight of bit vector STV.

I: Boolean identity matrix.

$L(G)$: linked list representation of G.

M: the number of single short faults given by $M = \#E$.

$M(C)$: color mapping of a set of nets C.

MRV: matrix of response vectors.

MTV: matrix of test vectors.

N: number of nets in the interconnect circuit given by $N = \#C$.

N_P: number of pins in an interconnect circuit.

N_b: number of possible synthetic behaviors of a short.

N_s: number of simulated stains.

N_{r_s}: number of sampled multiplicity r_s shorts.

O: the set of all open faults (obviously $\#O = N$).

$P(r)$: probability of fault multiplicity r.

$P_F(F)$: probability of fault F.

$P_{Fo}(F^{r_o})$: probability of open fault F^{r_s}.

$P_{Fs}(F^{r_s})$: probability of short fault F^{r_s}.

$P_b(b)$: probability of fault behavior b.

$P_o(r_o)$: probability of open fault multiplicity r_o.

$P_s(r_s)$: probability of short fault multiplicity r_s.

$PRV[t]$: parallel response vector received at instant t.

PS: pin spacing.

$PTV[t]$: parallel test vector applied at instant t.

PV: production volume.

R: maximum short radius.

S: set of shorted nets or subgraph of shorted nets ($S = (V, E'')$).

SFR: solder fault rate.

SOR: solder open rate.

SRV_U: serial response vector of the same responding set of nets U.

SRV_i: serial response vector received by the receiver of net n_i.

$SRV_i[t]$: bit received at instant t.

SSR: solder short rate.

STV_i: serial test vector applied by the driver of net n_i.

$STV_i[t]$: bit applied at instant t.

T: suspect subgraph or stain ($T = (U, E')$).

U: set of suspect nets.

V: faulty net subset.

W: Hamming weight of serial test vectors.

Y: production yield.

Y_o: opens' yield.

Y_s: shorts' yield.

b: a particular behavior of a short.

c_i: a particular color in graph G.

$d_{i,j}^{r_s,r_o}$: diagnosability of fault $F_{i,j}^{r_s,r_o}$.

d^{r_s}: average diagnosability of multiplicity r_s shorts.

$d^{r_s\prime}$: estimate of d^{r_s}

$d_i^{r_s}$: diagnosability of short $F_i^{r_s}$.

$[d_i^{r_s}]_b$: diagnosability of short $F_i^{r_s}$ given the behavior is b.

$e_{i,G}$: single short fault between net n_i and the ground net.

$e_{i,P}$: single short fault between net n_i and the power net.

$e_{i,j}$: single short fault between nets n_i and n_j.

f: a generic single fault.

k: number of vertex colors in graph G.

k_{min}: chromatic number of graph G.

m_T: number of faults that cause stain T.

n_G: the ground net.

n_P: the power net.

n_i: a particular net of C ($i \in \{1, 2, \dots, N\}$).

o_i: single open fault in net n_i.

p: serial test vector length in bits.

r: multiplicity of fault F ($r = r_s + r_o = \#F$).

r^{max}: practical maximum fault multiplicity.

r_s^{max}: practical maximum short multiplicity.

r_o: open multiplicity of fault F (# single opens).

r_s: short multiplicity of fault F (# single shorts).

r_s': short multiplicity of diagnosis F' (# single shorts).

t: discrete time for digital test sequence application ($t \in \{1, 2, \dots, p\}$).

α: clustering parameter.

α': estimate of α.

α_o: opens' clustering parameter.

α_s: shorts' clustering parameter.

ϵ: maximum DR error due to limiting fault multiplicity.

ϵ_s: maximum DR_s error due to limiting short multiplicity.

ε: maximum stain extent.

γ_o: probability that only open faults are present, given that the EA is faulty.

γ_s: probability that only short faults are present, given that the EA is faulty.

λ: average number of faulty solder joints per EA.

λ': estimate of λ.

λ_o: average number of single opens per EA.

λ_s: average number of single shorts per EA.

μ: estimate of fault multiplicity mean value.

σ: estimate of fault multiplicity standard deviation.

τ_F: time for locating the single faults in fault F.

$\tau_{F'}$: time for locating the single faults in diagnosis F'.

About the Authors

José T. de Sousa (http://esda.inesc.pt/~jts) is an assistant professor and a researcher at INESC/IST, Technical University of Lisbon. He received a BS and an MS degree in electrical and computer engineering from IST in 1989 and 1992, respectively, and a PhD degree in electronics from the Imperial College of Science, Technology and Medicine, University of London, in 1998. From 1998 to 1999 José was an invited postdoctoral researcher at Bell Laboratories in Murray Hill, New Jersey. His research interests include digital circuit design and test, reconfigurable computing, networking and multimedia systems.

Peter Y.K. Cheung (http://infoeng.ee.ic.ac.uk/~peterc) is a Reader in Digital Systems and Deputy Head of the Department of Electrical and Electronic Engineering at the Imperial College of Science, Technology and Medicine, University of London. He graduated from Imperial College in 1973 with first class honors and was awarded the IEE prize. After working for Hewlett Packard for a few years, he was appointed a lecturer at Imperial College in 1980. Peter runs an active research group in digital design, attracting support from many industrial partners. He was elected as one of the first Imperial College Teaching Fellows in 1994 in recognition of his innovative teaching. His research interests include VLSI architectures for signal processing, asynchronous systems, reconfigurable computing, behavioral modeling and test.

Index